Benvenuto, Plaumann (Eds.)
Green Chemistry in Industry
Green Chemical Processing

Green Chemical Processing

Edited by
Mark Anthony Benvenuto

Volume 3

Green Chemistry in Industry

Edited by
Mark Anthony Benvenuto and Heinz Plaumann

DE GRUYTER

Editors
Prof. Dr. Mark Anthony Benvenuto
University of Detroit Mercy
Department of Chemistry & Biochemistry
4001 W. McNichols Rd.
Detroit, MI 48221-3038
USA
benvenma@udmercy.edu

Dr. Heinz Plaumann
University of Detroit Mercy
Department of Chemistry & Biochemistry
4001 W. McNichols Rd.
Detroit, MI 48221-3038
USA
hmplaumann@gmail.com

ISBN 978-3-11-056113-5
e-ISBN (PDF) 978-3-11-056278-1
e-ISBN (EPUB) 978-3-11-056153-1

Library of Congress Control Number: 2018950024

Bibliographic information published by the Deutsche Nationalbibliothek
The Deutsche Nationalbibliothek lists this publication in the Deutsche Nationalbibliografie; detailed bibliographic data are available on the Internet at http://dnb.dnb.de.

Cover image: scyther5/iStock/Getty Images Plus
Typesetting: Integra Software Services Pvt. Ltd.
Printing and binding: CPI books GmbH, Leck

www.degruyter.com

About the series

Green Chemical Processing is a continuing series of volumes composed of refereed chapters, with upcoming volumes having submission dates of 15 June and 15 December each year. All areas of green chemistry, pending as well as established, are considered and welcome.

If you are interested in contributing a chapter, please contact series editor Mark Benvenuto, of the University of Detroit Mercy, at: benvenma@udmercy.edu concerning the appropriateness of your topic. We are interested in any and all new ideas that examine any of the twelve principles of green chemistry.

For more information on all previous and upcoming volumes of *Green Chemical Processing*, see www.degruyter.com/view/serial/462886.

https://doi.org/10.1515/9783110562781-201

Contents

List of contributing authors

Steven J. Bachofer
Department of Chemistry
Saint Mary's College of California
Moraga, CA 94575
USA
bachofer@stmarys-ca.edu

Paula Berton
Department of Chemistry
McGill University
801 Sherbrooke St. W.
Montreal, QC H3A 0B8
Canada
paula.berton@mcgill.ca

Agnieszka Gajewski
BASF Corporation
1609 Biddle Ave
Wyandotte, MI 48192
USA
agnieszka.gajewski@basf.com

Keith E. Gutowski
BASF Corporation
1609 Biddle Ave
Wyandotte, MI 48192
USA
keith.gutowski@basf.com

Philip G. Jessop
GreenCentre Canada
945 Princess Street
Kingston, ON K7L 0E9
Canada
philip.jessop@greencentrecanada.com

Steven P. Kelley
Department of Chemistry
McGill University
801 Sherbrooke St. W.
Montreal, QC H3A 0B8
Canada
steven.kelley@mcgill.ca

Pei-Yu Kuo
Department of Forestry and Natural Resources
National Ilan University
No. 1, Section 1, Shennong Road
Yilan City, Yilan County
Taiwan
peiyu.kuo@mail.utoronto.ca

Mark D. Lingwood
Department of Chemistry
Saint Mary's College of California
Moraga, CA 94575
USA
mdl4@stmarys-ca.edu

Jing Luo
Faculty of Forestry
University of Toronto
33 Willcocks Street
Toronto, ON M5S 3B3
Canada
luo.jing@mail.utoronto.ca

Andreas Metlen
AMT1
Molenakkersstraat 51
2910 Essen
Belgium
metlen@hotmail.com

Raymond Neff
BASF Corporation
1609 Biddle Ave
Wyandotte, MI 48192
USA
raymond.neff@basf.com

Laura M. Reyes
GreenCentre Canada
945 Princess Street
Kingston, ON K7L 0E9
Canada
laura.reyes@greencentrecanada.com

https://doi.org/10.1515/9783110562781-202

Robin D. Rogers
525 Solutions, Inc.
720 2nd Street
Tuscaloosa, AL 35403
USA
robin.rogers@mcgill.ca

Philip G. Sliva
Amway Corp
7575 Fulton Street East
Ada, MI 49355
USA
phil.sliva@amway.com

Nicole Tratnik
Faculty of Forestry
University of Toronto
33 Willcocks Street
Toronto, ON M5S 3B3
Canada
nicole.tratnik@mail.utoronto.ca

Ning Yan
Faculty of Forestry
University of Toronto
33 Willcocks Street
Toronto, ON M5S 3B3
Canada
ning.yan@utoronto.ca

Mark A. Benvenuto and Heinz Plaumann

1 Green chemical processes

Abstract: This volume includes chapters written by corporate chemists as well as academic chemists and covers a wide variety of subjects, including: the model of academic and corporate cooperation used by the Green Centre Canada; biomass treatment via catalysis; uses and potential uses of ionic liquids; means of greening the production of liquid rinse fabric softeners; how polyurethane production can be greened; the use of bark as a renewable source for higher value chemicals; aspirin production in the teaching lab, and how students can connect improvements in the synthesis to green chemical principles. The inclusion of several corporate authors in this volume represents an expansion of the overall green chemistry discussion and shows how the 12 principles of green chemistry can be incorporated into large-scale chemical processes.

Keywords: Industrial-scale green chemistry, Green Centre Canada, renewables

1.1 Introduction

Welcome to the third volume of the growing series *Green Chemical Processing*. As this series grows and expands, we are trying to continue to direct the evolution of it toward broad and inclusive discussions of numerous subjects and points of view when it comes to green chemistry and the improvement of chemical processes.

Included in this volume are excellent chapters written by corporate chemists from major companies, as well as others. Such authors have not been as prevalent in the literature of green chemistry as academic authors have been, largely because the mission of academic chemists and chemical engineers includes writing for the open literature and broadly for the profession, while that of corporate personnel may or may not, depending on the "climate" of a company. In broad terms, corporate chemists must by the nature of their job focus their energies on research and processes that will either become patents or that will be strictly guarded as trade secrets. Publication in the open literature is second to this.

And yet, the changes made in our world that will ultimately make it greener and more sustainable for the near and distant future are not those that qualify as proof-of-concept experiments and developments run solely in the laboratory. Rather, it is changes on an industrial scale that will translate to greener processes and products that will be seen by the public, and that indeed will be world-changing as they are implemented and become the norm. Even if a green change is something used but not thought of often, such as an improvement in the production of a commodity

https://doi.org/10.1515/9783110562781-001

Figure 1.1: Garbage truck that utilizes natural gas.

chemical like polyurethane, the improvement can make a very positive difference, while still going unnoticed. When an improvement is seen by the public, such as garbage trucks running on natural gas – seen in Figure 1.1 – as opposed to diesel, even though the change may be an incremental one (natural gas is still a fraction of petroleum refining), it places the discussion and awareness of green processes to the fore.

1.2 This volume

This volume has what we hope is a selection of chapters that will make readers think about how broadly the principles of green chemistry can be applied.

1.2.1 Green Centre Canada

We are flattered and honored to have Drs. Phil Jessop and Laura Reyes author a chapter in this volume about the activities of the Green Centre Canada. The Green Centre has become a model of how to nurture a concept which has one or more of the principles of green chemistry associated with it, and move it forward to some level of commercialization. The Centre is at least in part designed to help start-up companies and entrepreneurs bridge what is sometimes rather cynically called the "valley of death," meaning the point between the gestation of an idea and the point at which some product, material or service is brought to commercial viability.

Importantly, as this volume comes to publication, the timing corresponds to the 10th anniversary of the Green Centre Canada, which is under the expert tutelage of the very same Prof. Philip Jessop, who has produced numerous patents and seen a large number of papers written in association with the Centre. We believe this chapter offers a great deal to others, whether they are in academia, industry, or government.

1.2.2 Biomass dissolution and polyoxometallate catalysis

Catalysis is one of the central points of green chemistry, and the use of specific catalysts for the treatment of biomass becomes an especially good utilization of them. Professor Rogers' team has provided an excellent chapter on what is called the delignification of specific biomass using polyoxometallates. The potential for the large-scale use of such catalysts is intriguing, and this article seems to be one of only a few in this specific area.

1.2.3 Ionic liquids

Dr. Gutowski of BASF has contributed a chapter to this volume on the uses that ionic liquids have and may have in the future, in various large-scale chemical processes in industry. His survey is very thorough, and he does an excellent job of comparing the uses and advantages of ionic liquids to the costs involved (such as those of production of the liquid, as well as clean-up) when such processes move from the lab bench to industry.

1.2.4 Liquid rinse fabric softeners

Dr. Philip Sliva, at Amway, has provided a chapter on a class of materials that many people take for granted, or do not think of at all, yet that all of us ultimately must use: fabric softeners. The production of what get called consumer end-use items and materials is an ultimate aim of the entire chemical industry. Yet certain materials, fabric softeners included, are simply considered mature products or product lines for which no further research is needed and for which no greening is possible. Dr. Sliva proves this wrong with a well-thought out examination of how these materials are produced, how they are sourced, and what the challenges to their large-scale production will be in the future.

1.2.5 Renewable polyurethanes

Raymond Neff and Agnieszka Gajewski of BASF have contributed a chapter on one of the materials that have helped revolutionize many of the products we use in modern life, and indeed that has helped modern society achieve the quality of life enjoyed by many today: polyurethanes. It is probably fair to say that few in the general public have thought much about this class of materials, and even that few chemists and chemical engineers are well versed in the sourcing and production of polyurethanes. A list of what products are made from them is probably enormous, and we can wonder if anyone has even tried to compile an exhaustive one in recent years. But this large and expanding list means that currently the source material, petroleum, will be the limiting factor for future growth. This team discusses polyurethanes not only in terms of their current production and ultimate sourcing, but with an eye to how everything from the source material to the process can be made greener.

1.2.6 Bark in bio-based materials

A further chapter on an important topic thought of by few people, but in this case one with huge commercial potential, as well as what might be called "green potential," has been provided to us by Kuo, Yan, Tratnik, and Luo. The use of bark – a material often either discarded or burned as a low-grade fuel – to produce useful, valuable chemicals, and materials is precisely the sort of subject this series intends to address. It is a step in the right direction for a specific facet of a large, mature industry that will only increase in size as more people utilize the products of our managed forests.

1.2.7 Aspirin in the lab

We also have included in this volume a chapter on aspirin synthesis, a highly established procedure and a well-used product that has been produced on an industrial scale for over a century, yet that is still taught in the organic chemistry laboratory. Professors Bachofer and Lingwood bring new, green developments to this process, and describe how these are presented to their students.

1.3 Summary

The chapters in this third volume represent what we believe to be a widening of the discussion of the principles of green chemistry to new areas, and from purely academic voices to a mixture including corporate voices. As we have mentioned in previous volumes, the chapters collected here do not all fall neatly within what have become the “normal” green chemistry subjects. But expanding the discussion to include authors from the corporate world is imperative if we expect to make substantial change to greener processes at the industrial level.

We do recommend anyone who is interested in what we have presented here examine the first two volumes in this series, to see who has authored chapters in them, and what subjects have been covered. This series continues to expand, and we are very willing to speak with any potential authors who might be interested in writing for a future volume.

Philip G. Jessop and Laura M. Reyes

2 GreenCentre Canada: an experimental model for green chemistry commercialization

Abstract: Promising chemistry technologies are difficult to commercialize because of the "commercialization gap" that exists between academia and industry. This is especially important for discoveries in the area of green chemistry that can only fulfil their environmental and societal promise if they are successfully adopted by the chemical industry. However, the existing technology transfer model for academic commercialization is not well-suited for the highly sector-specific and long-term needs of chemistry technologies. GreenCentre Canada was founded in 2009 as a response to these commercialization needs: a chemistry-focused centre with sector-specific expertise (a Sector-specific Commercialization Centre, or SCC), including both highly trained scientists and business development professionals. GreenCentre works with academic researchers throughout Canada and internationally to evaluate, de-risk, scale-up, and optimize early-stage technologies in order to demonstrate the technology potential to industrial buyers or customers. Additionally, GreenCentre's work extends to small- and medium-sized enterprises at a more advanced stage in the technology development process, as well as large multinational enterprises that are well-established within the chemical industry but also benefit from the centre's expertise and resources. GreenCentre Canada represents a unique model for the development and commercialization of green chemistry technologies so that they may realize their environmental and societal benefits.

Keywords: green chemistry, commercialization, technology transfer, scale-up, innovation

2.1 Why do we need a new model for commercialization?

Green chemistry discoveries that never enter the marketplace do not help the environment. That harsh reality tends to make academic green chemistry researchers more applications-oriented than their peers in traditional fields like inorganic chemistry. That motivation to help the environment pushes many academic green chemists to choose projects that address commercial needs, but that well-meaning effort hits a brick wall after they invent something that has real market potential. That brick wall,

This article has previously been published in the journal Physical Sciences Reviews. Please cite as: Jessop, P. G., Reyes, L. M. GreenCentre Canada: An Experimental Model for Green Chemistry Commercialization. *Physical Sciences Reviews* [Online] **2018**, *3* (6). DOI: 10.1515/psr-2017-0189

https://doi.org/10.1515/9783110562781-002

sometimes described in terms of other metaphors such as the "commercialization gap" or the "valley of death", is the near-impossibility of getting a university-developed chemistry technology into the market.

"Valley of Death" may sound melodramatic, but the grim description is apt. Far too many commercially-viable academic technologies fail on the pathway from invention to first revenue. Why? First and foremost is the problem of the gap between academic capabilities and industrial expectations. If a company is to consider licensing and developing a new technology from academia, they need sufficient evidence that the technology has a decent chance to generate revenue. In other words, the technology must be de-risked sufficiently. Scale-up needs to be demonstrated. Performance for specific applications needs to be proven with key benchmark tests. Kilograms of samples need to be tested for properties, performance, and reproducibility. Unfortunately, chemistry professors and their students can only make grams of sample, not kilograms. They have no scale-up facilities, and typically have no knowledge of standard industrial benchmark tests. Graduate students are also reluctant to do significant optimization and scale-up work because that is not normally considered appropriate for thesis work in a chemistry department. Therefore, faculty lack the ability to take the technology to the state of readiness required for industry uptake. That inability to meet the minimum expectations of industry frustrates the commercialization hopes of many chemistry professors.

Despite the difficulties, there are a few shining examples of university green chemistry inventions that successfully jumped the gap. One recent example is the polyurethane production process that consumes waste CO_2, developed at RWTH Aachen University, being commercialized in Germany by Covestro [1]. Another is the mercury-capturing ionic liquid developed at Queen's University Belfast and now in use in Malaysia [2]. Unfortunately, these are exceptions rather than the rule.

Technology transfer offices (TTOs) exist at most universities, especially in North America, to help faculty bridge the commercialization gap. In our initial visits to Canadian universities, GreenCentre Canada representatives often encountered tension between chemistry faculty and their TTOs, even though both parties were willing to work toward the same goal. That tension was a result of frustrations from past attempts at commercialization and a gloomy outlook on the potential for future successes. Why do professors, even with the assistance of TTOs, have difficulties commercializing green chemistry technologies? If a new technology has a competitive edge over the incumbent technology, such as being cheaper or more effective, and it is better for health and the environment, then why should it be challenging to bring it to market? Part of the problem is a clash between the enormous effort required to commercialize chemical processes and the limited capabilities of the typical TTO. While the TTO model has had some key successes, it has inherent deficiencies with regards to its ability to move a chemical technology forward:

- TTOs are geographically-specialized rather than subject-specialized. In other words, they are dedicated to a single institution rather than a single subject like chemistry.
- TTO staff are rarely chemists (a consequence of the point above). Instead, they are required to cover all subjects in the university that may generate disclosures. Their role is as a generalist rather than a specialist [3] and therefore they lack the knowledge to evaluate the potential of chemistry-related disclosures.
- There are so many TTOs that companies looking for new chemical technologies choose to work with only a select few.
- Every university TTO has different policies and expectations on financial returns. Companies and venture capital sources often find the variety, rigidity, and excessive expectations of TTO's frustrating. For example, Siegel et al. [4] found that 80 % of corporate managers (but only 13 % of TTO directors) found universities to be "too aggressive in exercising intellectual property rights." Similarly, 80 % of managers but only 7 % of TTO directors felt that inflexibility of universities was a barrier to technology transfer.
- TTOs are not well-connected with either the venture capital community or the chemical industry because of the wide geographical distribution of TTOs and lack of subject specialization.
- TTOs have no labs for testing and development.
- TTOs have no scale-up facilities to de-risk scale-up, generate energy consumption data, or generate larger samples of a new material.
- TTOs do not have the financial resources to carry a technology through a multi-year development program.

Thus a new model is needed. The new model needs to resolve most if not all of the above problems. Fortunately, almost all of the above limitations are a direct consequence of a single feature of TTOs: their geographical rather than subject specialization. If one could have a new series of offices dedicated to commercialization (let us call them Subject-specific Commercialization Centres, or SCCs) to either supplement or replace the tradition TTOs, then those new offices could be subject specialized. For example, have a SCC for chemistry, another for electrical engineering, and yet another for medical devices. One for each field of endeavour. Each SCC would serve all of the universities in the country or the region rather than being dedicated to a single university. Each SCC would have staff trained in that subject and labs and facilities appropriate for the development and scale-up of technologies in that discipline. Such a new model would solve most of the problems inherent in the TTO model and would further the cause of green chemistry by more effectively facilitating the commercialization of green chemistry technologies from academia and entrepreneurs.

GreenCentre Canada was designed to be a test of the SCC model; it is therefore an experiment. Like any experiment being tried for the first time, it may succeed or fail, but in either case we will learn from the experience and hopefully inspire other countries to try the SCC model for the furthering of green chemistry

The benefits of the SCC model over the TTO model have been recognized for years [3, 5]. In other countries, similar tests of the SCC model have taken place, such as the UK's Medical Technologies IKC, a SCC for the commercialization of medical inventions [5].

Alternatives to the TTO and SCC models include other commercialization facilitators such as contract research firms, technology brokers, and technology incubators. Contract research firms, like SCC, are subject specialized but typically charge the university directly. Technology brokers are profit-making or not-for-profit organizations that help with the marketing and IP protection of a technology [6]. Like SCC and TTO, they often collect their reimbursement from license revenue rather than by charging a fee to the university. However, unlike SCC, they typically have no labs, and are therefore not capable of substantially contributing to the development of the technology. Examples include Research Corporation Technologies in the US and British Technology Group in the UK. Technology incubators assist start-up companies, rather than faculty, by providing office and, sometimes, lab space. Incubators associated with universities may provide access to university labs and researchers. Unlike SCC's, incubators typically charge fees, are not as specialized, and do not perform lab development or scale-up activities on behalf of the inventors.

2.2 Creating the subject-specific commercialization centre model

The vision to create GreenCentre Canada came from Dr. Rui Resendes, who joined PARTEQ, the TTO at Queen's University in Kingston, Canada in 2007. Quickly frustrated with the limitations of the TTO model with respect to chemistry and encouraged by his supervisor, John Molloy, he developed the SCC concept into a detailed plan, obtained the funding and launched the centre. The funding from the Canadian federal government, under the Centres of Excellence for Commercialization and Research program, started in January 2009, while the Ontario provincial government funding started in the summer of the same year. For the first year, the lab work took place in space borrowed from the Jessop group in the Department of Chemistry at Queen's University, by two scientists. In 2010 a new 10,000 ft^2 lab facility was opened in central Kingston, allowing GreenCentre to move out of the university and become an independent entity. The centre now has 28 staff, of which 16 are chemistry and chemical engineering researchers. In addition to the R&D lab in Kingston, the centre rents lab and office space, plus access to a 24,000 ft^2 scale-up and piloting facility, from the Xerox Research Centre of Canada in Mississauga, 3 hours west of Kingston.

Over the 8 years since the founding of GreenCentre, the centre has received 834 technology disclosures, including technologies from 305 researchers and 56 universities, most of them in Canada but with increasing numbers from universities outside the country such as Princeton and Oxford Universities. GreenCentre has issued 38 proof-of-principle awards to professors, filed 275 patent applications, been issued 27 patents, and completed 103 commercial innovation projects. The competencies covered by the staff include organic and organometallic synthesis, catalyst testing and screening, handling of air-sensitive/water-sensitive/pyrophoric materials, process scale-up and implementation, surfactant design and testing, polymer synthesis and characterization, application development and verification, and intellectual property protection.

2.2.1 Commercializing chemistry technologies

How does it work? How does GreenCentre help academics and entrepreneurs bring their inventions to the market? While GreenCentre also works with university-derived startup companies, here we will describe the process for academic technologies not yet part of a startup. When a professor at a university has an invention that she believes could be economically viable, the normal practice is to send an invention disclosure to the university's TTO. If an industrial licensee is immediately available, then there is no need for GreenCentre, but if, as in most cases, no licensee is forthcoming, then the TTO has the option to send the disclosure to GreenCentre Canada. The centre performs a preliminary screen, after which the technology will either be returned or will be licensed from the university for 90 days for a more complete assessment.

If that more extensive assessment is favourable, then the invention is "licensed in" to the Centre. GreenCentre will then undertake all of the work required to bring the technology to the point where it can be either licensed to an existing company or a spin-off company can be created, whichever seems more appropriate for that chemistry. The work can include lab development, testing against benchmark specifications, scale-up, patent protection, business case development, and marketing/outreach. Lab R&D is very focussed on questions that are crucial for market readiness. All of the considerable costs are borne by GreenCentre. If, at some point in the development, it becomes clear that the technology will not be viable after all, then the technology is returned to the university and GreenCentre's costs are not reimbursed.

If, however, the technology is licensed or spun out and then makes revenue, a previously-agreed portion of that revenue goes to the centre while the rest of the revenue goes back to the university and the inventors. Fortunately, because the centre is a not-for-profit organization, it does not need to do more than cover the costs of its day-to-day operations. The end result of the entire process, in successful cases, is a technology brought to market at no cost to the university (other than shared revenue). Back-ending the costs in this way makes the commercialization process affordable and less risky for universities, in contrast to having the development work performed by a contract research organization which would force the university to incur prohibitive costs and financial risks.

2.2.2 Advantages of the SCC model

The SCC model, as exemplified by GreenCentre Canada, has many advantages for the commercialization of green chemistry technologies:
- GreenCentre is subject-specialized (most of the staff are chemists or chemical engineers),
- There is only one chemistry SCC in Canada, so companies turn to GreenCentre for help in scouting new chemistry technologies,
- GreenCentre has dedicated staff who work to maintain connections with the chemical industry and the venture capital community,
- GreenCentre has labs for testing and development,
- GreenCentre has access to scale-up facilities to de-risk scale-up, generate energy consumption data, and generate larger samples of a new material, and
- GreenCentre, to some extent, has the financial resources to carry a technology through a multi-year development program.

Despite these many advantages, the SCC model is not without its challenges. Fortunately, these challenges can be overcome but only if the funding of the centre is designed with these issues in mind. The following four challenges are the key threats to the SCC model:
1. Because the SCC is field-specialized rather than geographically-specialized, the centre is far from most universities, so that communication with inventors and university TTOs is by phone or videoconference and rarely in person. Our experience shows that overcoming this challenge requires a concerted outreach effort including regular attendance at national chemistry conferences and visits to universities. This necessitates the dedication of significant financial and staff resources to outreach.
2. Because some chemical technologies require very expensive development programs, including the building and testing of demonstration plants, the costs of such development can be a severe strain on the resources of a small government-funded SCC.
3. While inventions in some other fields such as phone apps take only months to commercialize, the pathway to revenue for chemistry technologies typically ranges from 7 to 25 years. For that reason, license revenue is nearly guaranteed to be negligible in the first decade of the centre's existence. Thus a chemistry-specific SCC must rely on other sources of money to survive for at least the first 15–20 years. For this reason, GreenCentre receives government funding and does fee-for-service work for the chemical industry.

4. Because of the high cost of commercializing chemistry technologies, the SCC only has the resources to invest in a few. If those few technologies fail to become revenue-producing before the SCCs finances are drained, the centre's survival will be threatened. This can only partly be overcome by very judicious selection of the technologies to "license in" to the centre. Even then, successful outcomes for those technologies are far from guaranteed.

2.3 Working with academia, SMEs, and industry

At the core of GreenCentre's mission and operations is our work with academic researchers, following the motivation and SCC model described above. Additionally, we work with startups and small and medium-sized enterprises (SMEs) to similarly advance promising technologies within the green chemistry space. Lastly, we work with large multinational corporations, largely as partners for the development and commercialization of academic and SME technologies, or as fee-for-service providers. Although our approach in working with each of these three sectors is often similar, they all have different development needs, and we adapt our approach to best fit each client, technology, and project on a case-by-case basis.

In order to manage GreenCentre's cross-sector network and influx of technologies, an online portal was launched in 2015: Innovation House. Through this portal, GreenCentre receives all technology disclosures and accepts applications to active programs and funding opportunities.

2.3.1 Working with academic researchers

GreenCentre's work with academic researchers is integral to our mission. GreenCentre receives solicited and unsolicited early-stage technology disclosures that are ready to move beyond the basic research stage. Figure 2.1 shows the generalized summary of our commercialization process in working with academic technologies.

The process starts when a technology is disclosed to GreenCentre, usually by the researcher's associated TTO. A non-disclosure agreement must be in place with the institution, so that the technology may be reviewed in confidence. The technology disclosure documentation typically includes: published or draft manuscripts, intellectual property submissions, a brief description of potential benefits, and a non-confidential summary of the technology. At this stage, confirmation that the technology rights are accessible for commercialization is also required.

Upon receiving a technology disclosure, GreenCentre evaluates the technology through a quick screening. The quick screening will reveal if the technology is viable for a full assessment, or if it should be returned to the researcher without a more thorough analysis. Some factors considered at this stage include: technology risk and disruptiveness, commercial viability of patent position, fit within GreenCentre's

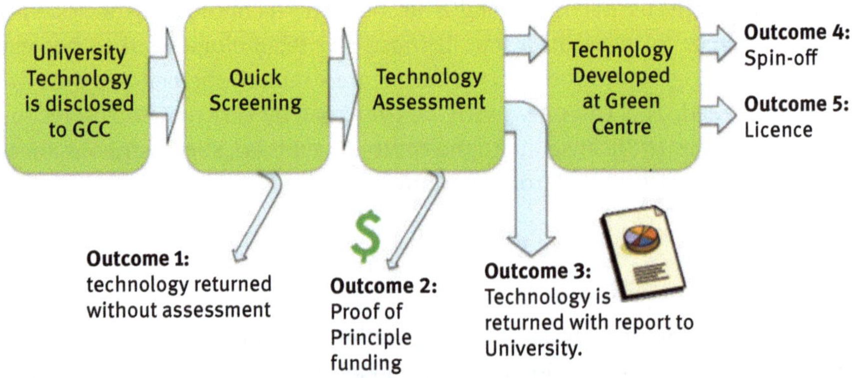

Figure 2.1: Summary of GreenCentre's process for evaluation and commercialization of chemistry technologies disclosed by academic researchers. At every stage and outcome possibility, the researcher and TTO are informed of internal and/or external feedback regarding the technology, providing valuable information for further development.

mandate and resources, path to market, technology cost and market estimate, regulatory concerns, and possible competition.

If a technology is deemed viable for a full technology assessment, a 90-day exclusive option to in-license the technology is secured, and a 1-page non-confidential "Technology Brief" summary is prepared by GreenCentre for circulation to select industry partners relevant to the technology. This is done to assess applications and potential markets, and gauge the level of interest from potential development partners or buyers. In addition, the technology is assessed more thoroughly from a technical and business perspective, including consideration of environmental impact, intellectual property, and product/process development.

The assessment includes economic viability and environmental impact compared to the incumbent technology. The process begins with confidential communication with the original inventor and with potential industry users. The industrial feedback is valuable in determining the commercial viability of the technology, the most appropriate market niche, the strength of the patent position, the price points that would be tolerable, the main competition, and the benchmark tests and specifications that the technology would need to meet. The centre also qualitatively assesses the likely environmental impact of the technology versus the incumbent technology. That environmental impact assessment is not nearly as rigorous as a life cycle analysis (LCA) because the technologies are at an extremely early stage in their development where there is insufficient data available for an LCA. Instead, the staff use a life cycle thinking approach involving qualitative predictions of whether the new technology, when scaled-up and adopted commercially, would be more, equally, or less environmentally damaging than the incumbent, on each of ten kinds of environmental impact (Table 2.1). If the disclosed technology looks promising from both financial *and* environmental perspectives, then GreenCentre licenses the technology on an exclusive basis.

Table 2.1: Rubric for life cycle thinking preliminary assessment of new discovery-stage technologies. For each risk category, the user estimates whether the new technology, as it would be applied at commercial scale after proper development and optimization, would be likely to be worse, roughly the same, or better than the incumbent technology.

Risk	Metric	New product or process	Existing product or process
Acidification	A risk if any compound is volatile AND contains N, S, F, Cl, or Br.		
Bioaccumulation	High risk if any compound has log K_{ow} > 4.3. Moderate risk if 3.5–4.3. Low risk if <3.5 or the compound is non-persistent.		
Eutrophication	High risk if any compound contains P or moderate risk if contains N. Decreased risk for very volatile compounds.		
Energy usage/ global warming	Does the process use a lot of energy? Does it emit N_2O, SF_6, or halogenated C_1-C_4 organics?		
Human toxicity	Moderate risk if any compound has an LD_{50} (rat, oral) of <500 mg/kg body weight. High risk if <50 mg/kg.		
Ozone depletion	A risk if any compound is volatile AND contains Br > Cl » F. (Except CH_2Cl_2 & $CHCl_3$ are not very problematic).		
Persistence	A risk if any organic compound has a high MW (e. g. polymers) or is highly halogenated or is a polycyclic aromatic.		
Resource depletion	A risk if large amounts of the following elements are used: Xe » He > Au > Ir, Rh, Ru > Kr, Os > Pt, Sb, Re, Hg, Se, Cd, Ne, Pd, Sn.		
Smog formation	A risk if any compound is volatile and organic.		
Other risks	Any other known risks, including chronic toxicity, airborne acute toxicity (by inhalation), ecotoxicity, flammability, explosion risks, endocrine disruption, etc.		

As a result of the technology assessment, a few outcomes may be reached: proof-of-principle funding is issued to the researcher to explore or address a specific issue raised by the assessment, the technology is returned to the institution with a full report including the application and market feedback from industry partners, or the technology is deemed viable for further development towards commercialization at GreenCentre.

These outcomes from the technology assessment are all beneficial to the researcher in different ways. For example, below is a testimonial from one academic researcher, Professor Natia Frank at the University of Victoria who works on light-induced magnetoresistive random access memory, for faster and more energy-efficient memory storage in computers and smartphones. Prof. Frank received a technology assessment report which allowed her to secure funding for her research, and provided her with access to key industrial partners to guide the applications of her work.

> "Working with GreenCentre has given me access to additional funding and industrial contacts that have greatly enhanced my research program. GreenCentre's assessment program alone provided commercial validation and was a key element in a successful NSERC I2I funding application. Also, they have arranged for research presentations to key industrial leaders, which has allowed me to engage in discussions with key industrial partners critical to guiding the research in directions pertinent to current industrial challenges." – *Prof. Natia Frank, University of Victoria*

If, following assessment, a technology is deemed viable for further development, the exclusive option is exercised and the technology is in-licensed by GreenCentre. Any technical and business development work necessary to get the technology to the stage where it is viable for market entry is carried out by GreenCentre, including:
- Application and product development
- Commercial benchmarking
- Scale-up of product or process
- Sustainability evaluation
- Intellectual property filings to protect technology
- Roadmap for market applications

In other words, GreenCentre undertakes all development work needed to bridge the commercialization gap between a promising academic technology to a market-ready technology that will be considered by industry. Additional to the technical and business development work itself, GreenCentre also engages strategic development partners from industry, which involves a commitment from these partners in terms of cash or in-kind contributions.

At the end of the development work, there are two possible outcomes for monetization of the technology: creating a spin-off company, or out-licensing the technology to industry. It is important to point out that it is only through the success of either of these two outcomes that GreenCentre receives any money back. At no other point during the academic commercialization model are costs recovered, as universities are not charged for any of the assessment or development work. GreenCentre instead uses capital acquired through federal, provincial, and fee-for-service funds to carry out the necessary work.

The most appropriate outcome – spin-off or out-license – is determined by considering the technology itself (product vs process), disruptiveness, potential market size, competition, etc. The adaptability of the GreenCentre model to choose the outcome best suited for each technology assures the best chance at successful commercialization and monetization. Some case studies of technologies and their paths to commercialization are outlined later in this paper.

2.3.2 Working with SMEs

GreenCentre's work with SMEs, similar to that with academic researchers, has typically been to develop relatively early-stage technologies. SMEs, especially early start-ups, often do not have all the resources needed to get their technologies through the commercialization gap. For example, a start-up founded from academic research is usually comprised of a highly technical team, and could lack business development experience and/or network reach into the chemical industry and application market. Even on the technical side, expertise is commonly lacking in process scale-up and optimization, or even simply lack of resources available such as equipment and time. Since GreenCentre is well-equipped in all of these aspects, SMEs can complement their existing strengths and leverage them against the centre's services and resources to more efficiently advance their technology's commercial adoption.

There have been two main ways in which GreenCentre works with SMEs: either as fee-for-service clients, or through GreenCentre-created programs that allocate government funding towards a development project, alleviating the costs from the SME client. There have been three such programs: the SME Commercialization Fund (SCF) program, the Ontario Network of Entrepreneurs (ONE) Chemistry program, and the InnovationHouse Chemistry Countering Climate Change (IHC_4) program. Each of these programs was funded by the provincial Government of Ontario, where GreenCentre is based, and thus focused on Ontario-based SMEs. The common goal of these programs has been to provide resources and services to SMEs towards a development project at no cost to the client (or at a reduced cost in the case of the SCF program), so that environmentally and societally beneficial technologies can be accelerated into the market.

The SCF program, operating in 2013, awarded successful SMEs with $50,000–150,000, up to 25 % of project costs, with the remaining 75 % secured by the client. The ONE Chemistry program, which operated from April 2015 to April 2016, accepted applications from SMEs with green chemistry technologies on a rolling basis. After an initial screening by GreenCentre for program eligibility, the applications and projects were brought before an external panel of experts on a monthly basis, who would then decide which projects would be worked on (similar to the assessment report process for academics).

More recently, the IHC_4 Program focuses on chemistry-based technologies that have the potential to reduce greenhouse gas (GHG) emissions, funded by climate change initiatives from the Ontario government. This program has been run so far as two distinct SME Competitions (rather than a rolling application process), the first on October 2016 and the second on June 2017. GreenCentre first reviews the applications that are submitted during the application window, assessing for eligibility criteria, technology viability, and fit within GreenCentre's capabilities. The culminating event is one day of back-to-back presentations, where the chosen finalists present their technology, business case, and development project directly to an external panel of experts, who rank and ultimately select the competition winners. The development projects which GreenCentre completes are valued at approximately $200,000 CAD per client. The format of the IHC_4 SME Competitions allows the company representatives themselves to pitch their technology directly to experts, acquiring valuable experience as well as immediate feedback that could help guide the company in its growth.

2.3.3 Working with large enterprises in industry

GreenCentre continuously engages industrial partners during the commercialization process for academic technologies, starting at the technology assessment stage and potentially including monetization through joint ventures or out-licensing. Partnerships with large multinational enterprises are therefore central to GreenCentre's work with early-stage technologies.

Additionally, GreenCentre performs fee-for-service (FFS) work for large enterprises, such as:

- Evaluation and design of new chemistries, chemical processes, and applications
- Optimization of current synthesis and processes
- Custom chemical and material synthesis
- Scale-up of chemical processes to pilot-scale
- Environmental analysis and sustainable process design and development

The FFS contracts with industry enable GreenCentre to leverage these funds in addition to government funding programs, which help support the work carried out with academics and SMEs. Therefore, the inclusion of FFS work helps GreenCentre be more financially sustainable, especially in its relative infancy before significant revenue is returned from commercialized technologies.

2.4 Green chemistry technologies at GreenCentre Canada

In the commercialization process for academic technologies, it is only at the final stage of monetization – either out-licensing to industry or creating a spin-off company – that revenue can be made. Therefore, GreenCentre must carefully evaluate and re-evaluate the commercialization potential of those technologies which are disclosed, assessed, and developed. For certain technologies, a partnership with already-established manufacturers/suppliers/distributors to sell a product is the easiest and most promising route for market entry. For others, more dedicated resources may be required and the creation of a separate spin-off company makes the most sense.

One example of the former commercialization route is an immobilized catalyst technology that was invented by Professor Steven Bergens and his group at the University of Alberta [7]. The Bergens catalyst is a rhodium-based transfer hydrogenation catalyst. Its potential comes from not only the catalyst itself, but also a polymer backbone that is covalently attached to the ligand. This polymer can be deposited on a solid support, effectively immobilizing the catalyst as a solid material without a significant change in its performance. In other words, the catalyst has the efficiency and enantioselectivity of a homogeneous system, but because it is immobilized on a solid support, it can be used as a heterogeneous catalyst for flow systems for high throughput and recyclability.

The catalyst-containing solid, after some preliminary validation performed by GreenCentre, was proposed to industry partners to gauge market interest and guide development work. Alongside this work, GreenCentre also engaged four industry leaders from the fine chemicals and pharmaceutical sectors as development partners, funding further work at GreenCentre as well as the Bergens research lab. Currently, the catalyst material is undergoing validation testing as a cartridge for flow reactors. If this testing proves successful, the cartridges will be manufactured and made available for purchase, with Chiral Technologies as a global distribution partner.

A very different technology and development story came from the Jessop lab at Queen's University. The technology uses switchable salts – an amine-based compound that readily switches between ionic and neutral states with the addition/removal of CO_2 gas [8]. The Jessop lab and GreenCentre found potential application of this technology in wastewater purification through forward osmosis [9, 10]. In essence, the amine salt draws out water from the wastewater stream, and clean water is recovered easily by CO_2 desorption and evaporation of the volatile amine, both of which can then be recycled back into the process.

In contrast to the Bergens catalyst cartridges, the Jessop forward osmosis technology is a process, one which requires demonstration of scale, performance, and potential applications. Therefore, this technology was spun out in 2012 as its own company – Forward Water Technologies (FWT). Currently, FWT is based at the Xerox Research Centre of Canada, GreenCentre's second location. Having access to Xerox facilities and scale-up engineers, FWT is operating a micro pilot unit within the pilot plant space to demonstrate performance at industrially relevant scales, as shown in Figure 2.2.

The two examples above, the Bergens catalyst and Forward Water Technologies, reflect the varying spectrum of technologies that are brought into and commercialized by GreenCentre. The Bergens catalyst is a product that benefits from development and distribution partners, requiring a relatively small amount of time and capital before it can be successfully brought to market. In contrast, FWT is a process that requires large-scale development with a dedicated scale-up pilot unit for application testing, in turn requiring a larger amount of time, capital, and resources before

Figure 2.2: The demonstration unit for wastewater purification, built by Forward Water Technologies in Mississauga, Canada.

exit. For an SCC like GreenCentre, a careful balance needs to be struck between the technology's market potential, and the resources that it might require between disclosure and monetization.

2.5 Lessons for the future

As has been outlined in this chapter, the SCC model that GreenCentre exemplifies is a highly beneficial and necessary one for the commercialization of chemistry technologies, especially for those under the green chemistry umbrella that can only enact their environmental and societal benefits if successfully implemented. There do still remain a few key challenges to address, however.

Since researchers submit disclosures to GreenCentre, the technologies that are brought in are limited to the areas that academia is working on. Those areas may not necessarily align with those technologies that have the biggest potential for commercialization or environmental impact. In fact, within the scope of publications that use the term "green chemistry", there is an inversely proportional relationship between the number of papers published in an area and its associated environmental impact. That means that the academic technologies that are disclosed are more likely to be ones that do not have much potential for disruptiveness or environmental benefit on a large scale. In order to help address this, GreenCentre has created a section on its InnovationHouse portal to list industry challenges. These challenges include industry-wide unsolved problems as well as promotion of active company-specific challenges, and both are listed to help guide researchers on potential routes for their work and to promote collaborations.

Another key challenge is that of funding. The initial funding period of 10 years granted to GreenCentre by the Government of Canada is less than the average timeline to get a chemistry-based technology out to market and start making revenue. Therefore, GreenCentre needs to remain creative in its fundraising in order to extend its presence past the point at which license revenue is brought in by the early technologies that were commercialized. This has already been put into practice by focusing on commercialization of more products than processes, as in the Bergens technology previously highlighted. Products require significantly less capital, time, and fewer resources to develop to the point of market entry, especially if development and monetization partners are engaged early on.

Despite these remaining challenges, the GreenCentre SCC model is continuously proving itself. Specialized centres like GreenCentre are more efficient at commercializing green chemistry technologies than the traditional TTO model, providing the necessary technical and business expertise, equipment and ability to scale-up, and network of industrial partners. Given the timeline of 7–25 years needed to monetize chemistry-based technologies to the point of revenue, the experiment of the GreenCentre model is well underway. The flexibility of GreenCentre to adapt its strategy based on the technology itself has proven to be quite valuable, since the incoming chemistry-based technologies fall under a wide range of potential applications and markets. Solving the problem of the commercialization gap or "valley of death" is crucial if green chemistry discoveries are to lead to a better tomorrow.

References

[1] Scott A. Learning to love CO2. Chem Eng News. 2015;93(45):10–16.
[2] Abai M, Atkins MP, Hassan A, Holbrey JD, Kuah Y, Nockemann P, et al. An ionic liquid process for mercury removal from natural gas. Dalton Trans. 2015;44:8617–24.
[3] Chapple W, Lockett A, Siegel D, Wright M. Assessing the relative performance of U.K. university technology transfer offices: parametric and non-parametric evidence. Res Policy. 2005;34:369–84.
[4] Siegel DS, Waldman D, Link A. Assessing the impact of organizational practices on the relative productivity of university technology transfer offices: an exploratory study. Res Policy. 2003;32:27–48.
[5] Williams CJ, Ellwood P, Roberts J, Thorpe R. Closing the translation gap – medical technologies innovation. In: The XXIV ISPIM Conference – Innovating in Global Markets: Challenges for Sustainable Growth; 16–19 June 2013, Helsinki, Finland: ISPIM, 2013.

[6] In: Schmoch U, Reid PP, Encarnacao J, Abramson HN, editor(s). Technology transfer systems in the United States and Germany: lessons and perspectives. Washington, DC: National Academies Press. National Academy of Engineering and Fraunhofer Institute for Systems and Innovation Research 1997.
[7] Bergens SH, Kalapugama S, Nepal P, McGinitie E. Catalyst systems for use in continuous flow reactors and methods of manufacture and use thereof. International Patent Application PCT/CA2014/050732.
[8] Mercer SM, Jessop PG. "Switchable water": aqueous solutions of switchable ionic strength. Chem Sus Chem. 2010;3:467–70.
[9] Jessop PG, Mercer S. Aqueous solvents with reversibly switchable ionic strength. Provisional U. S. Patent Application Patent 61/303,170.
[10] Jessop PG, Mercer S. Systems and methods for use of water with switchable ionic strength. International Patent application Patent PCT/CA2011/050777.

Steven P. Kelley, Paula Berton, Andreas Metlen
and Robin D. Rogers

3 Polyoxometalate catalysts for biomass dissolution: understanding and design

Abstract: The use of polyoxometalate catalysts for selective delignification of biomass presents a possible route toward using ionic liquids (ILs) to efficiently obtain high-molecular weight biopolymers from biomass. Rapid progress in this area will depend on recognizing and using the link with already well-developed inorganic chemistry in ILs pursued outside the field of biomass processing. Here, we use crystal structures determined from single crystal X-ray diffraction to better understand the behavior of $[PV_2Mo_{10}O_{40}]^{5-}$, a polyoxometalate catalyst known for its ability to promote selective delignification of biomass in the IL 1-ethyl-3-methylimidazolium acetate ($[C_2mim][OAc]$). The crystal structure of $[C_2mim]_5[PV_2Mo_{10}O_{40}]\cdot$ THF shows the formation of cationic shells around the anions which are likely representative of the interactions of this catalyst with $[C_2mim][OAc]$ itself. The reaction of NH_4VO_3 with $[C_2mim][OAc]$ is explored to better understand the chemistry of vanadium(V), which is critical to redox catalysis of $[PV_2Mo_{10}O_{40}]^{5-}$. This reaction gives crystals of $[C_2mim]_4[V_4O_{12}]$, showing that this IL forms discrete metavanadates which are obtained from aqueous solutions in a specific pH range and indicating that the basicity of $[OAc]^-$ dominates the speciation of vanadium (V) in this IL.

Keywords: Biomass, Catalysis, Inorganic chemistry, Ionic liquids, Polyoxometalate, Vanadium

3.1 Introduction

The discovery of cellulose dissolution in ionic liquids (ILs [1]) arrived at a time of great interest in the use of lignocellulosic biomass as a source of fuel and platform chemicals. Concerns about climate change and diminishing fossil fuel reserves have put a focus on converting biomass, especially from non-food crops, into fuel and, to a lesser extent, high value chemicals to offset the cost of the refinery [2]. Since then, ILs have been used to dissolve different types of biopolymers and biomass [3]. ILs are now heavily investigated for facilitating the chemical deconstruction of biomass, most often as biomass pretreatments [4] or reaction media for conversion of cellulose to small molecules [5–7].

This article has previously been published in the journal Physical Sciences Reviews. Please cite as: Kelley, S., Berton, P., Metlen, A., Rogers, R. Polyoxometalate Catalysts for Biomass Dissolution: Understanding and Design. *Physical Sciences Reviews* [Online] **2018**, *3*. DOI: 10.1515/psr-2017-0190

https://doi.org/10.1515/9783110562781-003

However, the ability to dissolve biopolymers in ILs has not been enough in and of itself for this technology to replace the entrenched fossil fuel economy.

We have instead envisioned the use of IL-based dissolution to overcome the need for synthetic chemicals altogether using them to make biopolymers from biomass directly into materials and products [8]. The physical dissolution of biopolymers from biomass sources by ILs leads to biopolymer solutions that can be formed with the same versatility as plastics. As there are no prior approaches which allow solution-based processing of biopolymers without chemical derivatization, IL-derived biopolymer materials have no competing technologies.

Biomass dissolution with the aim of producing materials faces a different set of challenges than chemical conversion. Materials research in this area is active but focused mainly on working with pure biopolymers such as microcrystalline cellulose [9, 10] or commercial yarns and fabrics [11, 12]. Materials such as fibers can be prepared directly from biomass dissolved in ILs but are often mechanically inferior due to the inclusion of other components of the biomass [13, 14]. Structural polymers in biomass are nearly always part of some sort of composite matrix which nature has evolved to prevent dissolution and chemical attack. Advanced materials can be made using high purity, high molecular weight biopolymers that are first extracted from biomass with an IL, isolated, and then re-dissolved to form the material [15, 16]. However, the ideal biopolymer separation would selectively extract biopolymers from biomass by breaking the chemical cross-links between components, allowing high-quality materials to be made directly from the extract with greater efficiency.

Lignocellulosic biomass is a quintessential case of this challenge, as the extraction of cellulose from it requires cleaving covalent bonds to the cross-linkers lignin and hemicellulose. This is the reason why a large range of ILs can dissolve cellulose, but only a few can dissolve the biomass itself [17]. Research on improving biomass dissolution in ILs has mostly focused on the solvent itself, such as the design of ILs with more basic anions for higher solubility [18] or the use of molecular cosolvents such as dimethyl sulfoxide [19]. More rapid dissolution can also be achieved by varying the process conditions, for instance conducting dissolution at above the glass transition temperature of lignin [20]. The most universal approach, however, might be to catalytically break the cellulose-lignin bonds, as this would open up the widest range of ILs and process conditions, allowing the process to focus on material preparation rather than biopolymer recovery.

The cleavage of cellulose-lignin bonds, which occurs in the extraction of cellulose by ILs, also happens during conventional recovery of cellulose by Kraft pulping. Techniques for improving the efficiency of pulping processes may therefore have applicability in ILs. In 2011, our group explored one such approach, the use of polyoxometalate (POM) catalysts to enhance the delignification of cellulose extracted using the IL 1-ethyl-3-methylimidazolium acetate ([C_2mim][OAc]) [21]. POMs were selected based on their use in catalyzing the oxidation of lignin in wood pulp bleaching using oxidizers such as O_2 and H_2O_2 instead of the conventional, more hazardous oxidizer Cl_2

[22]. It was found that the combination of POM catalysts and [C_2mim][OAc] resulted in delignified cellulose yields that were comparable or even superior to the industrial Kraft process but with a much simpler, less chemically intensive overall process.

While a wide variety of catalysts have been used to deconstruct biomass in ILs, they have still been mainly used in the context of converting biopolymers to small molecules [23]. Also, while the areas of biomass dissolution in ILs, inorganic chemistry in ILs, and metal-catalyzed oxidations are independently well-investigated, the links between the three are not often discussed. Here, we focus on the use of polyoxometalate catalysts to selectively delignify cellulose and aid in its extraction from biomass. The differences between catalysis in ILs and connections between other areas are discussed. We also present the crystal structure of a POM-based catalyst used in [C_2mim][OAc] and the crystallization of a vanadate salt from this IL. These structures serve as case studies to illustrate where fundamental experiments can aid in the basic understanding of these systems.

3.2 POM catalysts in IL-based biomass dissolution

POMs are polyatomic ions with at least three metal centers bridged into some kind of discrete cluster by oxygen atoms. The diversity of structures and metals is truly vast; examples range from the basic, trinuclear chromium(III) carboxylate clusters [24] to exclusively uranium-containing POMs with 60 metal centers [25]. The most common type used in catalysts are spherical clusters with the formula $[XM_{12}O_{40}]^{n-}$, known as Keggin anions, where X is a main group element such as phosphorous or silicon, and M is usually molybdenum(VI) [26]. Other metal ions can substitute Mo(VI) sites in the Keggin anion, and usually a certain number of vanadium(V) atoms are required for the ion to be active as an oxidation catalyst [27].

Reports on the use of POMs specifically to enhance the dissolution of lignocellulosic biomass in ILs are limited. The first was reported in 2011, where two forms of the POM $[PV_2Mo_{10}O_{40}]^{5-}$, its free acid form and a $[C_2mim]^+$ salt, were used to enhance the selective dissolution of cellulose from a commercially important softwood [21]. Ground and sieved wood dust was cooked in [C_2mim][OAc] at 110 °C for 8–16 h with varying amounts of $[PV_2Mo_{10}O_{40}]^{5-}$ in either form, and the cellulose-rich material was precipitated from this solution with 1:1 v/v acetone:water. The addition of $H_5PV_2Mo_{10}O_{40}$ resulted in faster dissolution, dissolution of coarser wood particles, and lower lignin content of the recovered cellulose. $[C_2mim]_4[HPV_2Mo_{10}O_{40}]$ led to even lower lignin content but also lower carbohydrate yields, indicating that this catalyst was more active but less selective.

The effects of $H_5PV_2Mo_{10}O_{40}$ on wood dissolution supported that cellulose-lignin bonds were cleaved selectively and that these bonds were responsible for inhibiting the selective dissolution of cellulose in the absence of $[PV_2Mo_{10}O_{40}]^{5-}$. This was further supported by the fact that $H_5PV_2Mo_{10}O_{40}$ was found to be less effective at delignification if it was added after the wood had been dissolved, as by this point many of the targeted bonds have already been cleaved. $H_5PV_2Mo_{10}O_{40}$ was also investigated in microwave-assisted dissolution of wood in [C_2mim][OAc] where it was found to be less effective than in thermal dissolution, likely because of the short dissolution time [28]. Further confirmation of lignin oxidation as part of the dissolution process was obtained by observing the effect of pure O_2 added as an oxidizer, which reduced the dissolution time from 8 to 6 h, reduced the yield of water-insoluble lignin, and led to the isolation of several extractable small-molecule lignin oxidation products [29].

There has also been a report on the use of a $[C_2mim]^+$ salt of $[Mo_{137}O_{372}(CH_3COO)_{30}(H_2O)_{72}]^{42-}$, a particularly large POM composed of 12 discrete spherical [Mo_{11}] clusters linked together [30], for the catalytic delignification of biomass [31]. In this case, delignification was performed on wood chips, rather than wood dust, by cooking in [C_2mim][OAc] followed by addition of NaOH. The biomass itself is not completely dissolved in this approach; rather, the cellulose is delignified as in conventional pulping. The addition of POM catalyst to the IL greatly increased delignification (7.0 % vs. 32 % lignin content of produced cellulose rich material vs. untreated wood, respectively), particularly at high POM loadings similar to those used in conventional pulping. As no additional reagents were added, delignification is presumably an oxidative process with O_2 from air as the oxidant. However, recyclability of this system was not investigated, so whether the POM acts as a catalyst or a promoter cannot be confirmed.

The interest in converting biomass to fuels has led to greater interest in the use of POMs for depolymerization of cellulose itself. In the context of materials preparation, understanding these reactions is still important so that they can be prevented. However, even in this area there are few studies on the use of ILs with actual biomass. In one case, ILs were made from Keggin anions $[PW_{12}O_{40}]^{3-}$ and $[SiW_{12}O_{40}]^{3-}$ and used directly as solvents for dissolution of biomass and conversion of cellulose to monosaccharides [32]. The POM $[Mo_8O_{26}]^{4-}$ has been patented for use in catalyzing hydrolysis of cellulose dissolved in IL to glucose [33]. Beyond these, most investigations of POM-catalyzed reactions directly on cellulose do not involve ILs as solvents [34], and those that use ILs more frequently involve model compounds such as cellobiose [35], sucrose or inulin [36], starch [37], and glucose [38].

Published information is limited and shows the adaptation of existing catalysts to ILs but not improvements to these catalysts based on the chemistry in ILs. Nevertheless, a few trends can be drawn from what is available. POMs can act as acid catalysts or oxidation catalysts, and both types of products have been obtained from carbohydrates. However, all reported delignification routes appear to be based on oxidative processes, so purely acidic POMs should be avoided to increase selectivity of delignification over cellulose hydrolysis. Regardless of whether cellulose is dissolved or not, delignification appears to be more selective when conducted on solid wood and should be integrated into the dissolution step. The deconstruction of cellulose in a neat POM-based IL indicates that excessive POM concentrations are destructive [32]. It is possible that selection of the correct IL or cellulose concentration could enable the use of greater POM loading for faster dissolution without cellulose de-polymerization if the IL can shield cellulose molecules from the catalyst. The strong hydrogen bond donor groups on cellulose that are far less abundant on lignin may offer a handle for such a chemical distinction.

3.3 Crystal structures of $[C_2mim]_5[PV_2Mo_{10}O_{40}]$ and $[C_2mim]_4[V_4O_{12}]$

3.3.1 Crystal structure of $[C_2mim]_5[PV_2Mo_{10}O_{40}]$

It is clear that the use of POMs for catalyzing biomass dissolution in ILs is a highly unexplored area, limited at present to only POMs dissolved in $[C_2mim][OAc]$. Even for aqueous systems, the mechanisms of these large, speciation variable catalysts are often poorly understood, and much more remains unknown about their interactions with IL ions. Furthermore, many of the POM catalysts reported for use in ILs are liquid themselves [32, 34], which is useful from an applied standpoint but further complicates their structural characterization.

Outside the area of biomass dissolution, however, inorganic chemistry in ILs is quite well developed, including the chemistry of POMs. A number of crystalline dialkylimidazolium salts of Keggin anions have been synthesized through similar routes to their aqueous preparation but using imidazolium ILs as solvents, such as $[C_2mim]_4[SiMo_{12}O_{40}]$ [39], $[C_2mim]_4[SiW_{12}O_{40}]$ [40], and $[C_4mim]_3[PMo_{12}O_{40}]$ ($[C_4mim]$ = 1-butyl-3-methylimidazolium) [41]. Reaction of $H_3PMo_{12}O_{40}$ with $[C_4mim]Br$ also gives $[C_4mim]_3[PMo_{12}O_{40}]$ [42] in an analogous route to the preparation of $[C_2mim][PV_2Mo_{10}O_{40}]$ from its acid [21]. Structurally characterized salts of transition metal substituted Keggin ions have also been synthesized in ILs, including $[C_1mim]_2[H(SiFeW_{11}O_{40})]$ ($[C_1mim]$ = 1,3-dimethylimidazolium) based on the iron-substituted tungstosilicate anion [43] and $[C_4mim]_9[O(PFeWO_{39})_2]$, which contains an anion composed of two similar iron-substituted clusters bridged by an oxo ligand [44]. Crystal structures of dialkylimidazolium POM salts have also been used to show how the cation can affect catalytic activity [45], as was observed in the case of $[C_2mim]_4[HPV_2Mo_{10}O_{40}]$ vs. $H_5PV_2Mo_{10}O_{40}$.

Because of the wide range of applications of POMs in ILs, their reactions toward biomass are not especially well-developed. Here we present the crystal structure of a tetrahydrofuran (THF) solvate of $[C_2mim]_5[PV_2Mo_{10}O_{40}]$ to better understand the link between the delignification chemistry of this system and the established structural chemistry of other POMs in ILs. Combining aqueous solutions of $H_5PV_2Mo_{10}O_{40}$ and $[C_2mim]Cl$ gives an orange powder which was determined by elemental analysis to have the composition $[C_2mim]_4[HPV_2Mo_{10}O_{40}]$ [21]. This water-insoluble powder was washed with water until the resulting washes were pH-neutral and Cl^- free. Dissolution in DMSO followed by vapor diffusion with tetrahydrofuran (THF) gave single crystals of $[C_2mim]_5[PV_2Mo_{10}O_{40}]$·THF.

$[C_2mim]_5[PV_2Mo_{10}O_{40}]$·THF was refined in the monoclinic space group $P2_1/n$ with Z' = 1. The structure contains five full occupancy $[C_2mim]^+$ ions, a disordered $[PV_2Mo_{10}O_{40}]^{5-}$ anion, and a THF molecule disordered over at least 3 positions (Figure 3.1). The presence of 5 cations also confirms that the anion is fully deprotonated, which may indicate that the $[C_2mim]_4[HPV_2Mo_{10}O_{40}]$ from which it was recrystallized either disproportionate in solution or is itself a mixture of the fully deprotonated POM and the free acid. The $[PV_2Mo_{10}O_{40}]^{5-}$ ion was modeled with the V and Mo atoms disordered over all metal sites. This type of disorder has been observed in every other reported crystal structure containing the $[PV_2Mo_{10}O_{40}]^{5-}$ ion [46–50]. The ratio of $[C_2mim]^+$ ions to $[PV_2Mo_{10}O_{40}]^{5-}$ confirms that both V atoms are in their +5 oxidation state and that, at least in the solid state, the $[C_2mim]^+$ ion is stable toward the fully oxidized POM. The disorder in the structure prevents a detailed analysis of its bonding, but the geometries of the cations and anions are typical of dialkylimidazolium and α-Keggin ions, respectively.

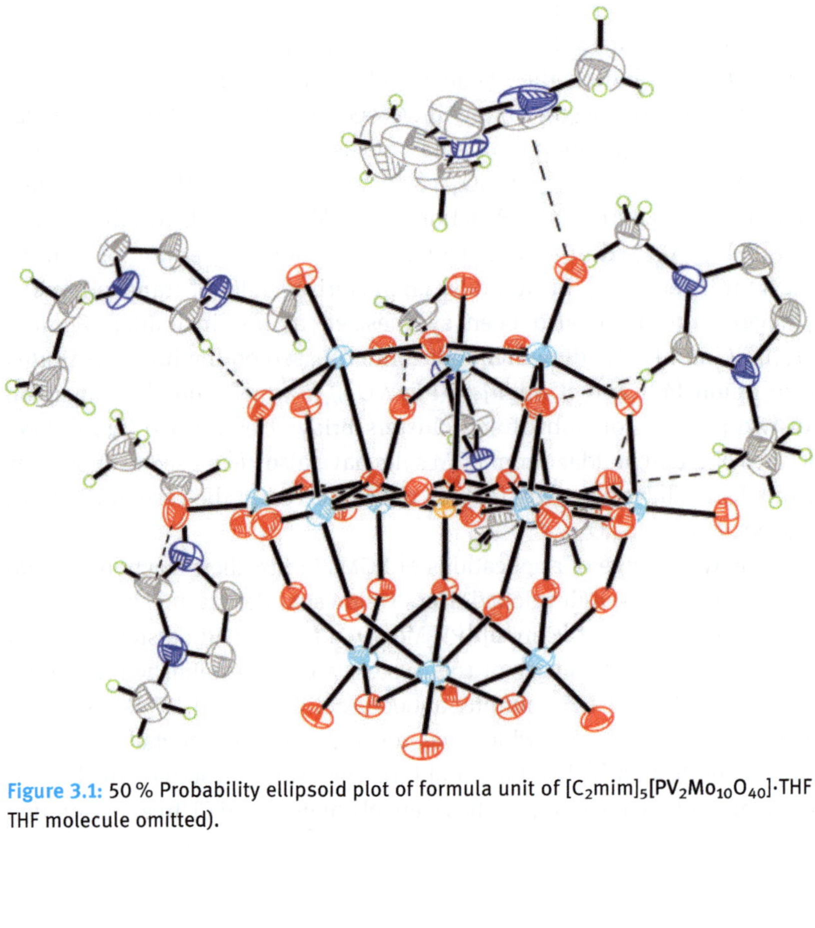

Figure 3.1: 50 % Probability ellipsoid plot of formula unit of $[C_2mim]_5[PV_2Mo_{10}O_{40}]$·THF (disordered THF molecule omitted).

Examination of the interactions between the $[C_2mim]^+$ and POM ions may give insight into the difference in catalytic activity between $[C_2mim]_5[PV_2Mo_{10}O_{40}]$ and the acid POM. It has been previously argued from crystallographic data that enhanced catalytic activity of a dialkylimidazolium POM salt could be due to the ability of the alkyl chains of the cation to stabilize and orient reaction intermediates [37]. Of the five symmetry-unique cations, three interact (make less than sum Van der Waals contacts) with neighboring anions through hydrogen bonding, one interacts through Coulombic interactions perpendicular to the plane of the ring, and one interacts through a mixture of both (Figure 3.2). The anion interacts with 15 cations, arranged around it in a roughly spherical shell. There are also short contacts between the anion and another anion, likely because they share a wall of this sphere, and between the

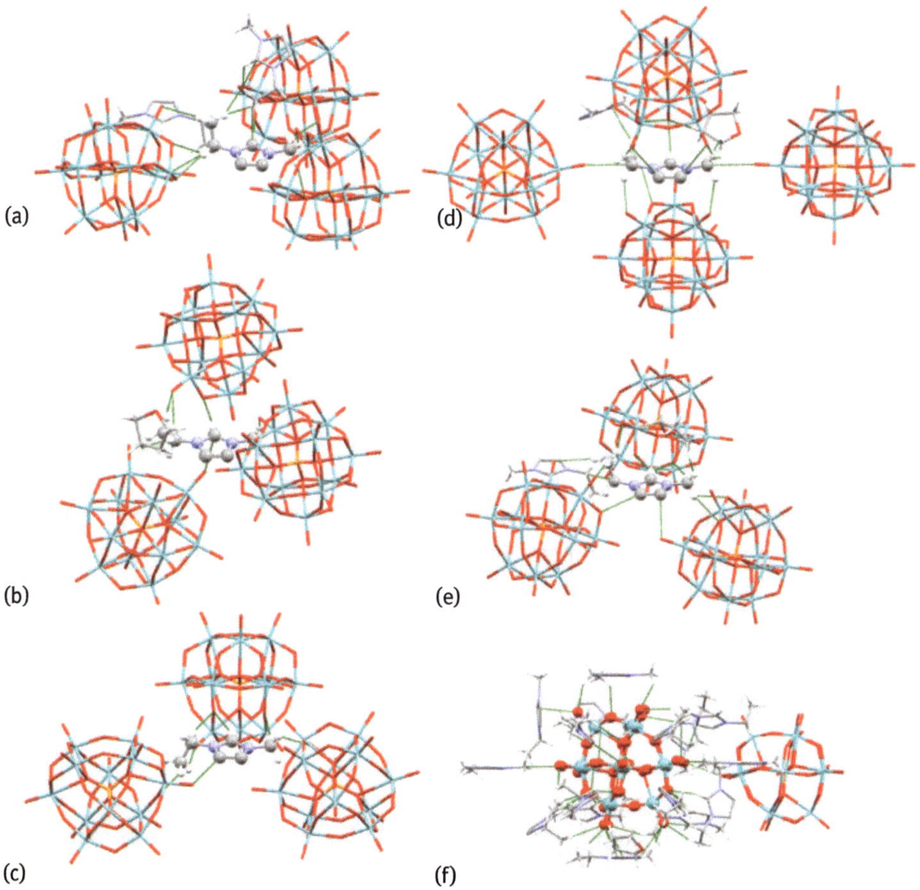

Figure 3.2: Packing diagrams showing intermolecular environments around cations interacting primarily through H-bonding (*a-c*), cations interacting primarily through Coulombic stacking interactions (*d* and *e*), and the anion (*f*). Dashed green lines indicate (less than the sum of the Van der Waals radii) contacts.

anion and two THF molecules. The THF molecule appears to fill void space between the alkyl chains of several cations and, given its disorder, probably does not engage in specific, strong interactions. The overall arrangement of ions resembles the close packing of spheres (Figure 3.3), which shows that electrostatic interactions (including hydrogen bonds) dominate the packing and indicates that $[PV_2Mo_{10}O_{40}]^{5-}$ is a good hydrogen bond acceptor.

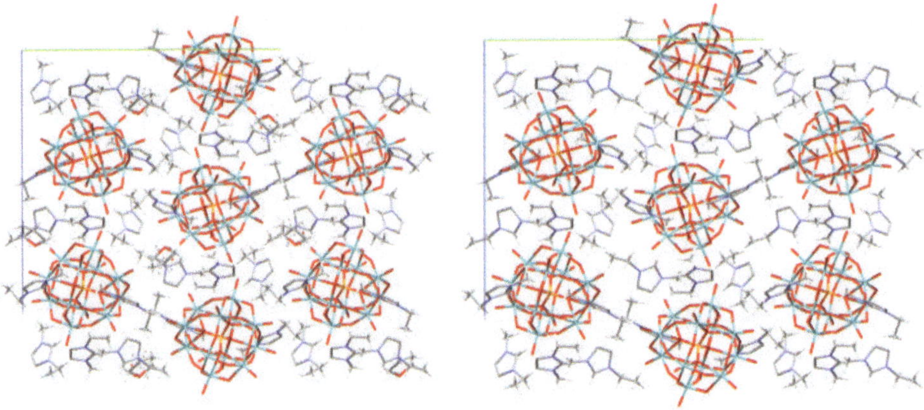

Figure 3.3: Packing diagram showing overall packing viewed down *a* axis (*left*, all atoms, *right*, THF molecules omitted). Axes are color coded: *b* = green, *c* = blue.

The increased catalytic activity and decreased selectivity of $[C_2mim]_5[PV_2Mo_{10}O_{40}]$ compared to its acid POM in $[C_2mim][OAc]$ is consistent with solvation by exclusively $[C_2mim]^+$ ions as observed in the crystal packing. For $[PV_2Mo_{10}O_{40}]^{5-}$ to participate in oxidation, an electron must be able to jump to it from its substrate. Such electron transfers are governed by Marcus theory, which states that their rate depends on the driving force of the electron transfer itself and the energy cost of reorganizing the solvent to accommodate the new species. $[C_2mim]_5[PV_2Mo_{10}O_{40}]$ in $[C_2mim][OAc]$, with an entirely cationic shell, would have a greater affinity for negatively charged electrons than $H_5PV_2Mo_{10}O_{40}$, which would very likely protonate and form hydrogen bonds with $[OAc]^-$ ions and thus have a mixed solvent shell.

3.3.2 Crystal structure of $[C_2mim]_4[V_4O_{12}]$

The crystal structure of $[C_2mim]_5[PV_2Mo_{10}O_{40}]{\cdot}THF$ does not give any information on the nature of the interactions between the IL and the vanadium atoms, since the vanadium and molybdenum sites cannot be distinguished. These sites are particularly important since vanadium imparts redox activity to the POM. As with IL-POM hybrid systems, vanadium chemistry in ILs has been well explored but generally not in the context of biomass. ILs are used to synthesize vanadium-containing inorganic solids [51, 52] and nanoparticles [53], as electrolytes for reversible batteries based on vanadium electrodes [54–56], and as solvents for vanadium-catalyzed reactions [57].

The basic chemistry of vanadium has been investigated for several reactive IL systems, but it is not particularly well developed for the non-reactive ILs and oxovanadate species that are more relevant to biomass dissolution. Mixed, anionic vanadium chloride species have been observed in chloroaluminate ILs as the result of direct reactions with vanadium(III) or (IV) chlorides [58] or deoxygenation of $VOCl_3$ by thionyl chloride [59], and V_2O_5 has been reacted with chloroaluminate ILs to give structurally characterized vanadium oxohalide compounds [60]. ILs containing vanadium oxifluoride ions have been prepared by reacting VOF_3 with imidazolium hydrogen fluorides, and the resulting dimeric $[(VOF_4)_2]^{2-}$ anions have been structurally characterized [61]. Metavanadate ($[VO_3]^-$) has been reportedly incorporated into ILs by reacting solutions of imidazolium or ammonium hydroxide salts with solutions of ammonium metavanadate (NH_4VO_3) [62]. A spectroscopic study has been published which examined the effects of ILs on aqueous solutions of vanadium, but the only observed effects on speciation were ascribed to chemical reactions with the anion [63].

Here, the reactivity of NH_4VO_3 with $[C_2mim][OAc]$ is discussed in order to complement the absence of information specific to vanadates in the crystal structure of $[C_2mim]_5[PV_2Mo_{10}O_{40}]$. The IL $[C_2mim][OAc]$ is also known to dissolve small organic molecules such as pharmaceuticals [64] or nucleobases [65] which are hard to dissolve in molecular solvents, as well as a few inorganic ammonium and alkali metal salts [66]. Although imidazolium salts of various anionic vanadium complexes are known, $[C_2mim][OAc]$ is particularly active toward oxidation reactions, for instance by elemental chalcogens [67], and could conceivably be more reactive than other ILs toward oxidation by a vanadium(V) oxoanion. NH_4VO_3 was observed to dissolve completely when combined with four molar equivalents of $[C_2mim][OAc]$ and heated to 90 °C without stirring. The solution was similar in color to that of an aqueous solution of NH_4VO_3. No further changes were observed in the solution after 72 h of continuous heating. When the solution was allowed to cool to room temperature, colorless crystals formed which were found to be highly deliquescent in ambient atmosphere. As NH_4VO_3 itself is stable in air, this confirmed the crystals to be a new compound.

The compound was identified by SCXRD as $[C_2mim]_4[V_4O_{12}]$ (Figure 3.4). The structure crystallizes in the monoclinic space group $P2_1/n$ with two unique cations and half of an anion per asymmetric unit. The $[V_4O_{12}]^{4-}$ anion is a cyclic tetramer composed of four corner-sharing VO_4 tetrahedra, forming an 8-membered ring. The conformation is grossly similar to a cyclohexane chair, with the four bridging O atoms and two of the V atoms occupying the vertices of the chair. The anion is almost entirely disordered over two positions. Both positions have the same conformation and are mostly overlapping; they differ mainly in the positions of some of the O atoms which change which of the two unique V atoms occupies a vertex of the chair.

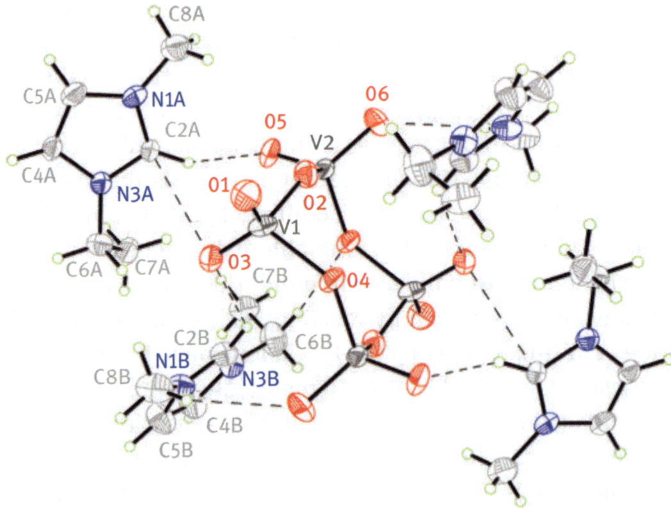

Figure 3.4: 50 % Probability ellipsoid plot of formula unit of $[C_2mim]_4[V_4O_{12}]$. Dashed lines indicate strongest interactions between ions. Unlabeled atoms are symmetry equivalents of labeled ones. Disorder omitted for clarity.

The cations show the typical geometry for $[C_2mim]^+$ with the ethyl group pointing out of the plane of the ring. One cation is entirely disordered over two slightly offset positions, while the other has disorder of the ethyl group over two close positions. The disorder of the cation appears to be a consequence of the disorder of the anion.

The anion in $[C_2mim]_4[V_4O_{12}]$ is a well-known form of metavanadate, and other cyclic, discrete forms of $[VO_3]^-$ have been observed in aqueous solutions at moderately alkaline pH [68]. Crystalline salts of $[V_4O_{12}]^{4-}$ have been obtained from basic aqueous solutions of vanadate [69, 70], hydrothermal reactions of V_2O_5 [71–73] or vanadate salts [74–77], aqueous reactions of vanadate salts with organic ligands [78, 79], and aqueous metathesis of NH_4VO_3 [80, 81]. However, the fact that the vanadium complex is known does not make its crystallization from $[C_2mim][OAc]$ trivially predictable. Vanadate speciation in aqueous solutions is particularly pH dependent, and for ILs, which always have acidic cations and basic anions at fixed relative concentrations, there is no scale analogous to pH which can be used to make such a prediction. Furthermore, Brønsted acids and bases (which include the vanadates) do not ionize in an IL to the same extent that they would in aqueous solution [82]. Thus, crystallization of $[V_4O_{12}]^{4-}$ from $[C_2mim][OAc]$ provides some of the first direct evidence that, for dissolved V^{5+} ions, the IL has comparable behavior to an aqueous solution at pH 8–9, and the basicity of the anion dominates. It is also notable that there are no prior reports of $[V_4O_{12}]^-$ being crystallized from a non-aqueous solution.

A compound with the same empirical formula, $[C_2mim][VO_3]$, has been reportedly isolated as an anhydrous, room temperature liquid by evaporation from an aqueous solution of NH_4VO_3 and $[C_2mim][OH]$ [62]. We also observed that if less than four equivalents of [C_2mim][OAc] were used, dissolution of NH_4VO_3 would occur but the liquids would vitrify rather than crystallize on cooling. The presence of other metavanadates at higher concentrations may frustrate crystallization in these systems.

Each anion makes short contacts (less than the sum of the Van der Waals radii) with 10 neighboring cations, six of cation A and four of cation B (as labeled in Figure 3.5). Cation A makes short contacts to three anions and four cations, all cation B. Cation B makes short contacts to two anions and six cations, four of cation A and two of cation B. Most of the cation-cation contacts are between C-H groups. They may be due to the fact that the highly charged anion pulls the cations close together or even simply to uncertainty in the positions of the atoms due to disorder. However, there are discrete pairs of cations stacking edge-over-edge, which may indicate $\pi \cdots \pi$ interactions. Cation A appears to interact more strongly with the anion, based on the greater number of cation-anion contacts and their nature. Cation A is complexed by two oxygen atoms from two different V centers. One accepts a C-H···O hydrogen bond, and the other engages in an electrostatic interaction with the C2 carbon atom. The remaining cation-anion interactions are hydrogen bonds from the hydrogen atoms around the cation ring or the C-H groups alpha to the nitrogen positions.

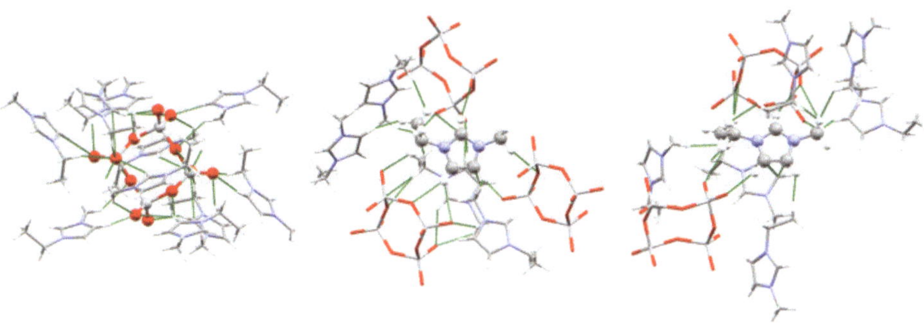

Figure 3.5: Short contact environments around $[V_4O_{12}]^{4-}$ (*left*), cation A (*center*), and cation B (*right*). Dashed green lines indicate less than sum VdW radii contacts.

$[V_4O_{12}]^{4-}$ can have multiple conformations, and it has been observed that the chair conformation is stabilized by forming supramolecular complexes with other molecules, such as water [70]. The cation in $[C_2mim]_4[V_4O_{12}]$ also appears to stabilize this form by offering two strong, positively charged sites of interaction at similar vectors to those of a water molecule. Although the larger polyoxometalates used in catalysis have more rigid structures and would not likely undergo such conformational changes, the observation of this effect here is an indicator of the strength of the interaction of a $[C_2mim]^+$ cation with a polyoxometalate.

The strength of the cation–anion interactions is also reflected in the packing, which can be conceived as a network composed of the anion and its two most strongly interacting cations, the two molecules of cation A which interact through the C2 position.

These units form a 3-D network of infinite 1-dimensional chains running orthogonally along both diagonals of the crystallographic *bc* plane (Figure 3.6). Hydrogen bonds between hydrogen atoms of the C4 and C5 positions of cation A and the anions of the neighboring orthogonal chains in both directions along *b* link these structures into an infinite network. Cation B reinforces this network through additional hydrogen bonding.

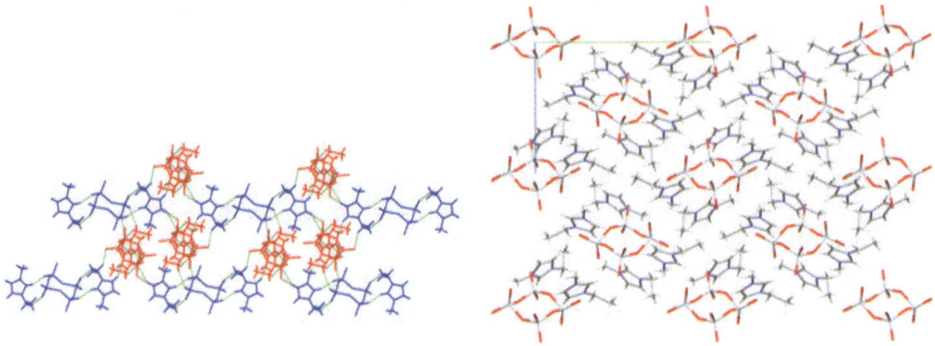

Figure 3.6: *Left*: Packing diagram showing the hydrogen-bonded network. Blue and red molecules are symmetry-equivalent but orthogonally-oriented 2:1 clusters of cation A and $[V_4O_{12}]^{4-}$. Green lines indicate hydrogen bonds. *Right*: 2x2x2 unit cell packing diagram viewed down *a*.

The crystal structure of $[C_2mim]_4[V_4O_{12}]$ offers further insight into the mechanism behind the effects of the counterions on the catalytic activity of $[PV_2Mo_{10}O_{40}]^{5-}$. Although the anion is structurally different from the infinite $[\{VO_3\}_n]^{n-}$ chains in NH_4VO_3 [83], it has the same empirical formula and can be formed through the following cation exchange reaction:

$$4[C_2mim][OAc] + 4NH_4VO_3 \rightarrow [C_2mim]_4[V_4O_{12}] + 4[NH_4][OAc]$$

It has been shown through spectroscopic studies on solutions of ammonium or alkali metal salts in [C$_2$mim][OAc] that the interactions of $[OAc]^-$ with the solute cation are the most important, and hydrogen bonding to $[NH_4]^+$ can be stronger than coordination to an alkali metal cation [84]. Thus, while NH_4VO_3 contains highly charged metal ions which could conceivably coordinate to $[OAc]^-$, they are still part of the anion, and dissolution of this compound in [C$_2$mim][OAc] is driven by solvation of the ammonium cation. The observation that hydrogen bonding can out-compete coordination to the metal cation of polyoxometalate is important, as anion-carbohydrate hydrogen bond formation is the mechanism of biomass dissolution in ILs [85].

3.4 Conclusions

The use of POMs as catalysts for biomass processing in ILs has emerged only recently and is limited to few examples. The area nevertheless shows promise, with both technical enhancements in performance and evidence for a selective cleavage of cellulose-lignin linkages having been discovered thus far. However, the current directions in research are limited to directly translating catalysts and techniques for other approaches into ILs. Both POMs and ILs are classes of compounds with extreme diversity, and design rules will be imperative in order to select the optimum performing systems. Studies from other areas of chemistry can be used here to contribute to a better understanding and design of catalysts for biomass processing.

POMs are also a challenging class of molecules to understand through structural characterization. The crystal structure of $[C_2mim]_5[PV_2Mo_{10}O_{40}]\cdot THF$ supports the identity of one such POM catalyst and shows how the interactions between the POM and IL ions may affect catalytic activity, but the structure of the POM itself is masked by crystallographic disorder that is inherent in this type of molecule. However, the smaller members of the class do provide an avenue for understanding the behavior of their functional groups in an IL. From the crystal structure of $[C_2mim]_4[V_4O_{12}]$, it can be seen that vanadates interact quite strongly with the IL cation and are shielded from the IL anion. More studies targeting such hybrid systems, rather than conventional inorganic solids or IL-based POMs, will be of benefit to biomass processing and to this area of inorganic chemistry at large.

Perhaps most importantly, the tool of catalysis links the separation of an extremely complex mixture, lignocellulosic biomass, to synthetic chemistry, which is arguably one of the most precise chemical sub-disciplines. The approach allows biomass to be reduced to its more easily understood bonds and functional groups. Presently, POM catalysts show promise for being a selective, efficient, and chemically non-intensive solution to the challenge of lignocellulosic biomass cracking, but even should they not turn out to be the ultimate solution, the outcome of these studies will be a better understanding of cellulose as a material and where it can be used most effectively in technology.

3.5 Experimental

The synthesis of $[C_2mim]_5[PV_2Mo_{10}O_{40}]$ has been previously reported [21]. NH_4VO_3 (Strem Chemicals Inc., Newburyport, MA) and $[C_2mim][OAc]$ (Iolitec USA, Tuscaloosa, AL) were used as received from their suppliers. $[C_2mim][OAc]$, a yellow-tinged liquid, was weighed into a tared 5 mL glass culture tube. The desired amount of NH_4VO_3, a tan powder, was calculated, weighed into a separate tared container, and then added at once to the $[C_2mim][OAc]$. The reactants were homogenized briefly with a glass stirring rod. The containers were sealed with Parafilm and then placed in a pre-heated sand bath at 90 °C. After 72 h, the heat was turned off, and the mixtures were allowed to cool slowly to room temperature. The reaction of NH_4VO_4 with four molar eq. $[C_2mim][OAc]$ produced crystals after 1 day. Reactions with 1, 2, or 3 eq. of $[C_2mim][OAc]$ remained transparent, vitreous solids at room temperature indefinitely. Crystals of $[C_2mim]_4[V_4O_{12}]$ were isolated for analysis by single crystal X-ray diffraction (SCXRD) by growing the crystals inside an evacuated, Drierite-filled desiccator, immediately covering them with Paratone upon removal, and rapidly isolating one and placing it on the diffractometer under a stream of cold N_2 gas.

The crystal structure of $[C_2mim]_5[PV_2Mo_{10}O_{40}]\cdot$THF was determined from data collected on a Bruker SMART diffractometer cooled under an LT-2 cryostat (Bruker AXS, Madison, WI). The structure was solved by direct methods and refined by full-matrix least squares refinement against F^2. Twelve unique metal atom sites (corresponding to the expected positions of metal atoms in a Keggin ion) were refined as containing both Mo and V with occupancies fixed at ratios of 5/6 and 1/6, respectively. V and Mo atoms occupying the same site were constrained to have identical positions and thermal parameters. For the disordered THF molecule, the atomic positions for three separate orientations were located and refined at fixed occupancies of 1/3. All three orientations were restrained to refine to have identical corresponding bond distances and angles, although the relative positions of atoms were freely refined. An additional restraint was applied to make the two C-O bond distances in the THF molecules equal. Thermal parameters were kept isotropic for the THF molecule. All other non-hydrogen atoms were refined anisotropically. Hydrogen atoms were placed in calculated positions and allowed to ride on the carrier atoms.

The structure of $[C_2mim]_5[PV_2Mo_{10}O_{40}]\cdot$THF contains a number artifacts which are likely connected with the disorder models. Such artifacts can indicate misinterpretation of a structure in some cases, but here they appear to indicate the true limit of reliable information in the data set. Several metal atom sites have large positive residual electron density peaks, which could be due to the fact that the two V atoms were refined as disordered evenly over all Mo sites but may preferentially occupy some sites or be in positions slightly offset from the Mo atoms. There are exceptionally short C-H···H-C contacts which may be due to unrefined positional disorder of the cations, which would itself be a consequence of the disorder in the anion and THF molecule. One $[C_2mim]^+$ ion has an unusually short C-C bond distance in the ethyl group which could be due to unrefined disorder or an artifact of unaccounted for electron density in the cluster. The correct identity of the structure is still supported by the fact that the ions refined to their well known, characteristic geometries and by the ratio of cations to anions, which is consistent with fully ionized $[PV_2Mo_{10}O_{40}]^{5-}$.

The crystal structure of $[C_2mim]_4[V_4O_{12}]$ was determined from data collected on a Bruker diffractometer equipped with a PLATFORM 3-circle goniometer and an Apex II CCD area detector. The crystal was cooled using an N-Helix cryostat (Oxford Cryosystems, Oxford, UK). The structure was solved by direct methods and refined by full-matrix least squares refinement against F^2. All non-hydrogen atoms were located from the difference map and refined anisotropically with a rigid bond restraint applied to all thermal parameters [86]. Disordered atoms were refined over two sites with occupancies fixed at 1/2. Hydrogen atoms were placed in calculated positions.

Methyl groups were refined using a riding-rotating model, and all other hydrogen atoms were constrained to ride on the carrier atoms.

Data integration and scaling were conducted with software packages from Bruker AXS (APEX 2, [87]). The Bruker SHELXTL software suite was used for structure solution and refinement and to draw probability ellipsoid plots [88]. Short contact analyses and packing plots were drawn using Mercury (Cambridge Crystallographic Data Center, Cambridge, UK) [89]. Structural data is available free of charge from the CSD (deposition numbers 1525882-1525883).

Acknowledgements: We would like to thank Dr. Mark Nieuwenhuyzen (Queen's University Belfast) for structure refinement of $[C_2mim]_5[PV_2Mo_{10}O_{40}]\cdot$THF. This research was initiated at the Queen's University Ionic Liquid Laboratory research center at Queen's University Belfast and continued in the chemistry departments of The University of Alabama and McGill University. This research was undertaken, in part, thanks to funding from the Canada Excellence Research Chairs Program.

References

[1] Swatloski RP, Spear SK, Holbrey JD, Rogers RD. Dissolution of cellulose with ionic liquids. J Am Chem Soc. 2002;124:4974–75.
[2] Cherubini F. The biorefinery concept: using biomass instead of oil for producing energy and chemicals. Energy Convers Manage. 2010;51:1412–21.
[3] Wang H, Gurau G, Rogers RD. Dissolution of biomass using ionic liquids. In: Zhang S, Wang J, Lu X, Zhou Q, editor(s). Structures and Interactions of Ionic Liquids Structure and Bonding Vol. 151. Heidelberg, Germany: Springer Berlin, 2013:79–105.
[4] Brandt A, Gräsvik J, Hallett JP, Welton T. Deconstruction of lignocellulosic biomass with ionic liquids. Green Chem. 2013;15:550–83.
[5] Rinaldi R, Palkovits R, Scuth F. Depolymerization of cellulose using solid catalysts in ionic liquids. Angew Chem Int Ed. 2008;47:8047–50.
[6] Wang P, Yu H, Zhan S, Wang S. Catalytic hydrolysis of lignocellulosic biomass into 5-hydroxymethylfurfural in ionic liquid. Bioresour Technol. 2011;102:4179–83.
[7] Zhang Y, Du H, Qian X, Chen EY-X. Ionic liquid-water mixtures: enhanced K_w for efficient cellulosic biomass conversion. Energy Fuels. 2010;24:2410–17.
[8] Rogers RD. Eliminating the need for chemistry. Chem Eng News. 2015;93:42–43.
[9] Rosewald M, Hou FY, Mututuvari T, Harkins AL, Tran CD. Cellulose-chitosan-keratin composite materials: synthesis, immunological and antibacterial properties. ECS Trans. 2014;64: 499–505.
[10] Sweely KD, Fox ET, Brown EK, Haverhals LM, De Long HC, Trulove PC. Inkjet printing ionic liquids for the fabrication of surface structures on biopolymer substrates. ECS Trans. 2014;64:575–82.
[11] Asaadi S, Hummel M, Hellsten S, et al. Renewable high-performance fibers from the chemical recycling of cotton waste utilizing an ionic liquid. ChemSusChem. 2016;9:3250–58.
[12] Haverhals LM, Reichert WM, De Long HC, Trulove PC. Natural fiber welding. Macromol Mater Eng. 2010;295:425–30.

[13] Sun N, Li W, Stoner B, Jiang X, Lu X, Rogers RD. Composite fibers spun directly from solutions of raw lignocellulosic biomass dissolved in ionic liquids. Green Chem. 2011;13: 1158–61.

[14] Barber PS, Kelley SP, Griggs CS, Wallace S, Rogers RD. Surface modification of ionic liquid-spun chitin fibers for the extraction of uranium from seawater: seeking the strength of chitin and the chemical functionality of chitosan. Green Chem. 2014;16:1828–36.

[15] King CA, Shamshina JL, Gurau G, Berton P, Khan NFAF, Rogers RD. A platform for more sustainable chitin films from an ionic liquid process. Green Chem. 2017;19:117–26.

[16] Shen X, Shamshina JL, Berton P, et al. Comparison of hydrogels prepared with ionic-liquid-isolated vs commercial chitin and cellulose. ACS Sust Chem Eng. 2016;4:471–80.

[17] Wang H, Gurau G, Rogers RD. Ionic liquid processing of cellulose. Chem Soc Rev. 2012;41: 1519–37.

[18] Abe M, Ohno H. Maintenance-free cellulose solvents based on onium hydroxides. Kagaku to Kogyo. 2010;63:891–93.

[19] Xu S, Zhang J, He A, Li J, Zhang H, Han CC. Electrospinning of native cellulose from nonvolatile solvent system. Polymer. 2008;49:2911–17.

[20] Li W, Sun N, Stoner B, Jiang X, Lu X, Rogers RD. Rapid dissolution of lignocellulosic biomass in ionic liquids using temperatures above the glass transition of lignin. Green Chem. 2011;13:2038–47.

[21] Sun N, Jiang X, Maxim ML, Metlen A, Rogers RD. Use of polyoxometalate catalysts in ionic liquids to enhance the dissolution and delignification of woody biomass. ChemSusChem. 2011;4:65–73.

[22] Gellerstedt G. Pulping Chemistry. In: Hon DN-S, Shiraishi N, editor(s). Wood and Cellulosic Chemistry, 2nd ed. New York and Basel: Marcel Dekker. CRC Press, New York 2001: 859–905.

[23] Wang H, Block LE, Rogers RD. Catalytic conversion of biomass in ionic liquids. In: Hardacre C, Parvulescu V, editor(s). Catalytic Conversion of Biomass in Ionic Liquids Catalysis Series. Cambridge, UK: RSC Publishing, 2014:1–19.

[24] Harton A, Nagi MK, Glass MM, Junk PC, Atwood JM, Vincent JB. Synthesis and characterization of symmetric and unsymmetric oxo-bridged trinuclear chromium benzoate complexes: crystal and molecular structure of $[Cr_3O(O_2CPh)_6(py)_3]ClO_4$. Inorg Chim Acta. 1994;217:171–79.

[25] Burns PC. Nanoscale uranium-based cage clusters inspired by uranium mineralogy. Mineral Magazine 2011, 75, 1–25.

[26] Keggin JF. The structure and formula of 12-phosphotungstic acid. Proc Royal Soc. 1934;144: 75–76

[27] Kozhevnikov IV. Catalysis by heteropoly acids and multicomponent polyoxometalates in liquid-phase reactions. Chem Rev. 1998;98:171–98.

[28] Wang H, Maxim ML, Gurau G, Rogers RD. Microwave-assisted dissolution and delignification of wood in 1-ethyl-3-methylimidazolium acetate. Bioresour Technol. 2013;136:739–42.

[29] Cheng F, Wang H, Rogers RD. Oxygen enhances polyoxometalate-based catalytic dissolution and delignification of woody biomass in ionic liquids. ACS Sust Chem Eng. 2014;2:2859–65.

[30] Müller A, Krickmeyer E, Bögge H, Schmidtmann M, Peters F. Organizational forms of matter: an inorganic super fullerene and keplerate based on molybdenum oxide. Angew Chem Int Ed. 1998;37:3359–63.

[31] Ozdokur KV, Moniruzzaman M, Yanik J, Ono T. Synthesis and characterization of a polyoxometalate-based ionic liquid catalyst for delignification of wood biomass. Wood Sci Technol. 2016;50:1213–26.

[32] Abia JA, Ozer R. Development of polyoxometalate-ionic liquid compounds for processing cellulosic biomass. Bioresour. 2013;8:2924–26.

[33] Wang J, Zhou M, Wang W, Yuan Y, Fu N. CN 105618139 A2016. A method for degradation of lignocellulose based on molybdenum polyoxometalate catalyst. June1.
[34] Hu L, Lin L, Wu Z, Zhou S, Liu S. Chemocatalytic hydrolysis of cellulose into glucose over solid acid catalysts. Appl Cat B. 2015;174-175:225–43.
[35] Li K, Bai L, Amaniampong PN, Jia X, Lee J-M, Yang Y. One-pot transformation of cellobiose to formic acid and levulinic acid over ionic-liquid-based polyoxometalate hybrids. ChemSusChem. 2014;7:2670–77.
[36] Chen J, Zhao G, Chen L. Efficient production of 5-hydroxymethylfurfural and alkyl levulinate from biomass carbohydrate using ionic liquid-based polyoxometalate salts. RSC Adv. 2014;4:4194–202.
[37] Chen X, Souvanhthong B, Wang H, Zheng H, Wang X, Huo M. Polyoxometalate-based ionic liquid as thermoregulated and environmentally friendly catalyst for starch oxidation. Appl Cat B: Environ. 2013;138-139:161–66.
[38] Chidambaram M, Bell AT. A two-step approach for the catalytic conversion of glucose to 2,5-dimethylfuran in ionic liquids. Green Chem. 2010;12:1253–62.
[39] Lin S, Liu W, Li Y, Wu Q, Wang E, Zhang Z. Preparation of polyoxometalates in ionic liquids by ionothermal synthesis. Dalton Trans. 2010;39:1740–44.
[40] Liu W, Tan H, Chen W, Li Y, Wang E. Ionothermal synthesis and characterization of two new heteropolytungstates with 1-ethyl-3-methylimidazolium bromide ionic liquid solvent. J Coord Chem. 2010;63:1833–43.
[41] Linguito SL, Zhang X, Padmanabhan M, et al. New polyoxomolybdate compounds synthesized in situ using ionic liquid 1-butyl-3-methylimidazolium tetrafluoroborate as green solvent. New J Chem. 2013;37:2894–901.
[42] Rao GR, Rajkumar T, Varghese B. Synthesis and characterization of 1-butyl-3-methyl imidazolium phosphomolybdate molecular salt. Solid State Sci. 2009;11:36–42.
[43] Chen W-L, Chen B-W, Tan H-Q, Li Y-G, Wang Y-H, Wang E-B. Ionothermal syntheses of three transition-metal-containing polyoxotungstate hybrids exhibiting the photocatalytic and electrocatalytic properties. J Solid State Chem. 2010;183:310–21.
[44] Santos FM, Brandão P, Félix V, et al. Organic-inorganic hybrid materials based on iron(III)-polyoxotungstates and 1-butyl-3-methylimidazolium cations. Dalton Trans. 2012;41: 12145–55.
[45] Chen D, Sahasrabudhe A, Wang P, Dasgupta A, Yuan R, Roy S. Synthesis and properties of a novel quaternized imidazolium $[\alpha\text{-}PW_{12}O_{40}]^{3-}$ salt as a recoverable photo-polymerization catalyst. Dalton Trans. 2013;42:10587–96.
[46] Plotnikova TE, Grigoriev MS, Fedoseev AM, Antipin MY. Crystal structure of praseodymium(III) complex $[Pr(DMSO)_8]_2PMo_{10}V_2O_{40}(NO_3)\cdot DMSO$ (DMSO = dimethyl sulfoxide). Russ J Coord Chem. 2004;30:60–67.
[47] Liu Y, Liu S, Liu S et al. Facile synthesis of a nanocrystalline metal-organic framework impregnated with a phosphovanadomolybdate and its remarkable catalytic performance in ultradeep oxidative desulfurization. Chem Cat Chem. 2013;5:3086–91.
[48] Liu X, Wang L, Yin X, Huang R. Assembly of hybrid materials based on a lanthanide-organic framework and a Keggin-type $[PMo_{12-x}V_xO_{40}]^{(3+x)-}$ (x = 1, 2) cluster. Eur J Inorg Chem. 2013;2013:2181–7.
[49] Huang X, Zhang X, Zhang D et al. Binary Pd-polyoxometalates and isolation of a ternary Pd-V-polyoxometalate active species for selective aerobic oxidation of alcohols. Chem Eur J. 2014;20:2557–64.
[50] Kalyani V, Satyanarayana VSV, Singh V, et al. New polyoxometalates containing hybrid polymers and their potential for nano-patterning. Chem Eur J. 2014;21:2250–58.

[51] Aidoudi FH, Aldous DW, Goff RJ, et al. An ionothermally prepared S = 1/2 vanadium oxyfluoride Kagome lattice. Nat Chem. 2011;3:801–06.

[52] Aidoudi FH, Black C, Athukorala Arachchige KS, Slawin AMZ, Morris RE, Lightfoot P. Structural diversity in hybrid vanadium(IV) oxyfluorides based on a common building block. Dalton Trans. 2014;43:568–75.

[53] Fechler N, Fellinger T-P, Antonietti M. Template-free one-pot synthesis of porous binary and ternary metal nitride@ N-doped carbon composites from ionic liquids. Chem Mater. 2012;24:713–19.

[54] Chou S-L, Wang J-Z, Sun J, et al. High capacity, safety, and enhanced cyclability of lithium metal battery using a V_2O_5 nanomaterial cathode and room temperature ionic liquid electrolyte. Chem Mater. 2008;20:7044–51.

[55] Jayaprakash N, Das SK, Archer LA. The rechargeable aluminum-ion battery. Chem Commun. 2011;47:12610–2.

[56] Plashnitsa LS, Kobayashi E, Noguchi Y, Okada S, Yamaki J. Performance of NASICON symmetric cell with ionic liquid electrolyte. J Electrochem Soc. 2010;157:A536–43.

[57] Conte V, Fabbianesi F, Floris B, et al. Vanadium-catalyzed, microwave-assisted oxidations with H_2O_2 in ionic liquids. Pure Appl Chem. 2009;81:1265–77.

[58] Hanz KR, Riechel L. Vanadium complexes in a Lewis basic room-temperature 1-ethyl-3-methyl-1H-imidazolium chloride/aluminum chloride molten salt. Inorg Chem. 1997;36:4024–28.

[59] Dent AJ, Lees A, Lewis RJ, Welton T. Vanadium chloride and chloride oxide complexes in an ambient-temperature ionic liquid. The first use of bis(trichloromethyl) carbonate as a substitute for phosgene in an inorganic system. J Chem Soc Dalton Trans. 1996;1996:2787–92.

[60] Mahjoor P, Latturner SE. Synthesis and structural characterization of $[bpyr]_4[V_4O_4Cl_{12}]$ and $[bpyr]_4[Bi_4Cl_{16}]$ grown in ionic liquid $[bpyr][AlCl_4]$ (bpyr = 1-Butylpyridinium). Cryst Growth Des. 2009;9:1385–89.

[61] Kanatani T, Matsumoto K, Hagiwara R. Syntheses and physicochemical properties of low-melting salts based on VOF_4^- and $MoOF_5^-$, and the molecular geometries of the dimeric $(VOF_4^-)_2$ and $Mo_2O_4F_6^{2-}$ anions. Eur J Inorg Chem. 2010;2010:1049–55.

[62] Branco A, Belchior J, Branco LC, Pina F. Intrinsically electrochromic ionic liquids based on vanadium oxides: illustrating liquid electrochromic cells. RSC Adv. 2013;3:25627–30.

[63] Bányai I, Conte V, Pettersson L, Silvagni A. On the nature of V(V) species in hydrophilic ionic liquids: A spectroscopic approach. Eur J Inorg Chem. 2008;2008:5373–81.

[64] Wang H, Gurau G, Kelley SP, Myerson AS, Rogers RD. Hydrophobic *vs.* hydrophilic ionic liquid separations strategies in support of continuous pharmaceutical manufacturing. RSC Adv. 2013;3:10019–26.

[65] Araujo JMM, Ferreira R, Marrucho IM, Rebelo LPN. Solvation of nucleobases in 1,3-dialkylimidazolium acetate ionic liquids: NMR spectroscopy insights into the dissolution mechanism. J Phys Chem B. 2011;115:10739–49.

[66] Pereiro AB, Araújo JMM, Oliveira FS, et al. Inorganic salts in purely ionic liquid media: the development of high ionicity ionic liquids (HIILs). Chem Commun. 2012;48:3656–58.

[67] Rodríguez H, Gurau G, Holbrey JD, Rogers RD. Reaction of elemental chalcogens with imidazolium acetates to yield imidazole-2-chalcogenones: direct evidence for ionic liquids as protocarbenes. Chem Commun. 2011;47:3222–24.

[68] Livage J. Vanadium pentoxide gels. Chem Mater. 1991;3:578–93.

[69] Roman P, San Jose A, Luque A, Gutierrez-Zorrilla JM. Observation of a novel cyclic tetrametavanadate anion isolated from aqueous solution. Inorg Chem. 1993;32:775–76.

[70] Nakano H, Ozeki T, Yagasaki A. $(Et_4N)_4[V_4O_{12}]\cdot 2H_2O$. Acta Cryst. 2002;C58:m464–5.

[71] Zhang Y, Zapf PJ, Meyer LM, Haushalter RC, Zubieta J. Polyoxoanion coordination chemistry: synthesis and characterization of the heterometallic, hexanuclear clusters

[{Zn(bipy)$_2$}$_2$V$_4$O$_{12}$], [{Zn(phen)$_2$}$_2$V$_4$O$_{12}$].H$_2$O, and [{Ni(bipy)$_2$}$_2$Mo$_4$O$_{14}$]. Inorg Chem. 1997;36:2159–65.
[72] Yang GY, Gao D-W, Chen Y, Jia H-Q. [Ni(C$_{10}$H$_8$N$_2$)$_3$]$_2$[V$_4$O$_{12}$].11H$_2$O. Acta Cryst. 1998;C54:616–8.
[73] Yi Z-H, Cui X-B, Zhang X, et al. Hydrothermal syntheses and structural characterizations of organic-inorganic hybrid materials of the M(II)-ligand/vanadium oxide system (M(II) = Mn(II), Cu(II) and Zn(II); ligand = 2,2'-bipyridine and 1,10-phenanthroline). Dalton Trans. 2007;2007:2115–20.
[74] Qi Y, Wang Y, Li H, et al. Hydrothermal syntheses and crystal structures of bimetallic cluster complexes [{Cd(phen)$_2$}$_2$V$_4$O$_{12}$]·5H$_2$O and [Ni(phen)$_3$]$_2$[V$_4$O$_{12}$]·17.5H$_2$O. J Mol Struct. 2003;650:123–29.
[75] Huang M-H, Bi L-H, Dong S-J. Bis[tris(2,2'-bipyridyl-κ^2*N,N'*)iron(II)] *cyclo*-tetravanadate decahydrate. Acta Cryst. 2004;E60:m153–5.
[76] Zhang X, You W-S, Zhu Z, Dang L, Sun Z, Zheng Z. Hydrothermal synthesis and characterization of a novel crystal containing [Co$_4$O$_4$]$^{4+}$ cubane core: [Co$_4$O$_4$(dpaH)$_4$(CH$_3$COO)$_2$]$_2$V$_4$O$_{12}$ · 5H$_2$O. Inorg Chem Commun. 2006;9:526–8.
[77] Tabatabaee M, Ahadiat G, Molčanov K. Tetra-kis(2-amino-4-methyl-pyridinium) cyclo-tetra-μ(2)-oxido-tetra-kis-[dioxido-vanadate(V)] tetrahydrate. Acta Cryst. 2011;E67:m1090.
[78] Žúrková L, Kucsera R, Gyepes R, Sivák M. Synthesis and X-ray crystal structure of two novel complexes: [M^{II}(*phen*)$_3$]$_2$V$_4$O$_{12}$·*phen*·22H$_2$O(M^{II}=Co, Ni; *phen*=phenanthroline). Monatsh Chem. 2003;134:1071–79.
[79] Morgenstern B, Steinhauser S, Hegetschweiler K. Complex formation of vanadium(IV) with 1,3,5-triamino-1,3,5-trideoxy-cis-inositol and related ligands. Inorg Chem. 2004;43:3116–26.
[80] Sharma RP, Singh A, Venugopalan P, Dansby-Sparks R, Xue Z-L, Rosetti S, et al. Stabilization of tetrameric metavanadate ion by tris(1,10-phenanthroline)cobalt(III): synthesis, spectroscopic and X-ray structural study of [Co(phen)$_3$]$_3$(V$_4$O$_{12}$)$_2$Cl·27H$_2$O. J Coord Chem. 2010;63:3016–27.
[81] Zhang K, Liang -D-D, Wang M-H, Luan GY. Bis[tris(2,2′-bipyridyl-κ^2*N, N′*)cobalt(II)] *cyclo*-tetravanadate undecahydrate. Acta Cryst. 2013;C69:138–41.
[82] Miran MS, Kinoshita H, Yasuda T, Susan MABH, Watanabe M. Physicochemical properties determined by ΔpKa for protic ionic liquids based on an organic super-strong base with various Brønsted acids. Phys Chem Chem Phys. 2012;14:5178–86.
[83] Syneček V, Hanic F. The crystal structure of ammonium metavanadate. Cechoslovackij Fiziceskij Zurnal. 1954;4:120–29.
[84] Oliviera FS, Cabrita EJ, Todorovic S, et al. Mixtures of the 1-ethyl-3-methylimidazolium acetate ionic liquid with different inorganic salts: insights into their interactions. Phys Chem Chem Phys. 2016;18:2756–66.
[85] Remsing RC, Swatloski RP, Rogers RD, Moyna G. Mechanism of cellulose dissolution in the ionic liquid 1-n-butyl-3-methylimidazolium chloride: A 13C and 35/37Cl NMR relaxation study on model systems. Chem Commun. 2006;12:1271–73.
[86] Thorn A, Dittrich B, Sheldrick GM. Enhanced rigid bond restraints. Acta Cryst Sect A Found Crystallogr. 2012;68:448–51.
[87] APEX 2 AXScale and SAINT, version. Madison, WI: Bruker AXS, Inc.; 2010.
[88] Sheldrick GM. SHELXTL, structure determination software suite, v.6.10. Madison, WI: Bruker AXS Inc; 2001.
[89] Macrae CF, Bruno IJ, Chisholm JA, Wood PA, Mercury CSD. 2.0 – new features for the visualization and investigation of crystal structures. J Appl Cryst. 2008;41:466–70.

Keith E. Gutowski

4 Industrial uses and applications of ionic liquids

Abstract: Ionic liquids are salts that melt at low temperatures (usually defined as less than 100 °C) and have a number of interesting properties that make them useful for industrial applications. Typical ionic liquid properties include high thermal stabilities, negligible vapor pressures, wide liquidus ranges, broad electrochemical windows, and unique solvation properties. Furthermore, the potential combinations of cations and anions provide nearly unlimited chemical tunability. This article will describe the diverse industrial uses of ionic liquids and how their unique properties are leveraged, with examples ranging from chemical processing to consumer packaged goods.

Keywords: ionic liquids, acid scavenging, hydrosilylation, hydroformylation, cellulose, electroplating, nuclear processing, gas compression, gas capture, dye-sensitized solar cell, battery, antistatic agent, consumer packaged goods

4.1 Ionicliquids

Ionic liquids as a class of materials have experienced a pronounced period of growth and exploration over the past three decades. However, prior to that these materials were far from new. The molten salt community continuously debates the origin or discovery ionic liquids, with people referencing Gabriel and Weiner's [1] 1888 observation of ethanolammonium nitrate (MP = 52 °C) as the first ionic liquid, with Walden's [2] 1914 ethylammonium nitrate discovery being the first "low-temperature" or "room temperature" ionic liquid (MP = 12 °C). This is arguably just a question of perspective. Sodium chloride, heated to a temperature of greater than 800 °C [3], is also an ionic liquid, albeit a high temperature ionic liquid, but is more commonly or aptly referred to as a molten salt. This discovery obviously predates the aforementioned examples. Furthermore, eutectic mixtures of quaternary ammonium salts with hydrogen bond donors, such as a 1:2 molar ratio of choline chloride and urea, have also been referred to as ionic liquids, but are more commonly called deep eutectic solvents [4]. Non-continuous phases of ionic liquids, or their use as additives in formulations and the like, often still retain the moniker of "ionic liquid" in the literature, but are simply salt solutions. From these examples, it should be obvious that the loosely named field of ionic liquids is rich with examples and diversity.

This article has previously been published in the journal Physical Sciences Reviews. Please cite as: Gutowski, K. Industrial Uses and Applications of Ionic Liquids. *Physical Sciences Reviews* [Online] **2017**, *3* (5). DOI: 10.1515/psr-2017-0191

https://doi.org/10.1515/9783110562781-004

Putting perspective aside, there is little argument that ionic liquids are a scientifically interesting class of materials with countless proposed uses and potential applications. This article will attempt to review some of these uses and applications as they relate to the chemical industry.

In general, ionic liquids are organic salts with melting points arbitrarily chosen to be less than 100 °C. To refine the definition further, room temperature ionic liquids, which are more suitable for many benchtop or ambient applications – and as the name implies – are organic salts with melting points below about 25 °C. Low melting organic salts are typically achieved using a variety of strategies, including utilizing bulky or asymmetric cations and/or anions with the intention to frustrate crystalline packing and by utilizing weakly-coordinating anions with diffuse or delocalized charges to weaken directional non-covalent interactions. With suitable choice of anions, ionic liquids can be water miscible or immiscible [5]. Wilkes and Zaworotko [6] are usually credited with preparing the first ionic liquids with broad diversity and applicability for air- and water-stable applications. Prior to 1992, the well-studied chloroaluminate ionic liquids [7] exhibited acute water sensitivity, limiting their widespread use, particularly in an industrial setting. The materials of Wilkes and Zaworotko were similarly based on imidazolium cations, but with air and water stable anions, including $[BF_4]^-$, $[NO_3]^-$, and $[CH_3CO_2]^-$ and were considered for uses in synthesis and electrochemistry. Fast forwarding to the present, the often referred to "infinite" number of cation–anion combinations has certainly grown exponentially, with cations spanning (but not limited to) alkyl ammonium, phosphonium, imidazolium, pyridinium and anions covering (but also not limited to) halides, sulfates, alkylsulfates, hexafluorophosphate, triflate, bistrifylimide, tosylate, and dicyanamide [8]. This chapter will cover ionic liquids of the Wilkes and Zaworotko type.

4.2 Chemical processing

Many early studies with ionic liquids focused on their used as solvents for performing chemical transformations, including biochemical transformations [9]. In this section, several non-exhaustive examples of the use of ionic liquids as enabling solvents for chemical processing are given, including reaction media for acid scavenging, addition reactions, and catalytic transformations.

4.2.1 Acid scavenging

BASF's patented [10] BASIL process is probably the most widely known and recognized example of industrial implementation of an ionic liquid-based process. BASIL stands for biphasic acid-scavenging utilizing ionic liquids [11]. As elucidated from the acronym, this utilizes an ionic liquid for acid scavenging in the production of alkylphenylphosphines, important precursors in the production of

BASF photoinitiators for UV curing of double bond-containing resins, like those used in coatings and printing inks.

The BASIL process is commercially important because of the significant process improvements realized versus the prior process. In the latter, acid scavenging in the reaction was achieved using alkylamines, like triethylamine, forming the protonated trialkylammonium halide salt. This product is a thick, dense, insoluble slurry that is difficult to handle and remove from the reactor, resulting in inefficiencies and the need for a processing solvent. By utilizing the ionic liquid precursor, 1-methylimidazole, as an alternative to triethylamine, the acid reaction forms 1-methylimidazolium chloride, a protic ionic liquid with a low melting point (75 °C) which separates and forms a biphasic system in the reaction vessel. The lower phase being comprised of pure ionic liquid can readily be removed, resulting in easy isolation of the upper phase product. Furthermore, the ionic liquid can be recycled by deprotonation, regenerating the 1-methylimidazole reactant. The schematic of the BASIL reaction is shown in Figure 4.1.

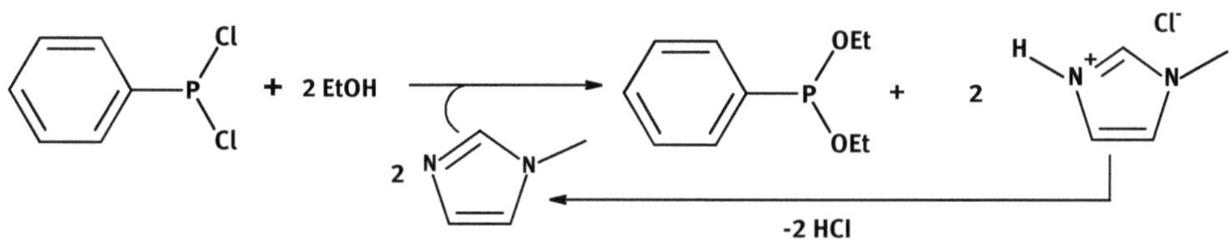

Figure 4.1: The BASF BASIL process.

The advantages of this new process are multifold. It is amenable to small reactor vessels like BASF's jet stream reactor. By eliminating solids, the reaction rate is improved, allowing for a practical multi-ton scale process. It is efficient, resulting in a 80,000-fold space-time yield increase to 690,000 kg m^{-3} h^{-1}. And it is easily extendable to other processes involving acid scavenging in a generic sense.

4.2.2 Hydrosilylation

Like BASF, Goldschmidt-Degussa (now Evonik) has been actively involved in extending ionic liquid use to industrial processes [12]. The most frequently referenced of these processes is the use of ionic liquids in the hydrosilylation reaction [13]. The hydrosilylation process transforms silanes and siloxanes (like polydimethylsiloxane) into organo-modified versions by the catalytic addition of the Si-H group to the C = C double bond of olefins (Figure 4.2). This is the most attractive and widely utilized method for large-scale production of organosilicon compounds because it is clean and highly efficient. However, the historical process is not without drawbacks, the main one being the inability to effectively recover the expensive platinum catalyst. Therefore, attempts are seldom made.

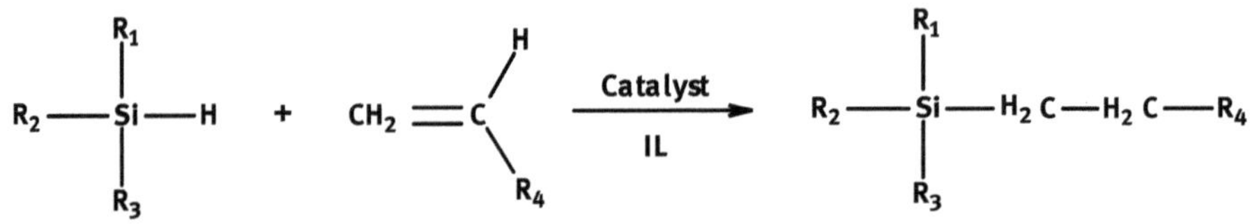

Figure 4.2: Hydrosilylation reaction.

A liquid–liquid biphasic approach has been developed by Degussa, taking advantage of the unique properties of ionic liquids to enable the process. The platinum catalyst is dispersed into the ionic liquid phase and the pure reaction product separates into a new phase upon completion. The product phase can easily be isolated and the ionic liquid phase containing the catalyst can be recycled and reused multiple times. The effectiveness of the process is improved with as low as 20 ppm catalyst loadings being possible, and leaching of the catalyst into the product phase is below the detection limits of 1 ppm. Favored ionic liquids include [C_4pyridinium][BF_4] and [C_4-3-piccolinium][BF_4]. Evonik has operated the ionic liquid-enabled hydrosilylation reaction on a pilot scale level, but no further information is available for a production-scale process.

4.2.3 Hydroformylation

Evonik also has a patented process [14] on the hydroformylation of olefins by the addition of syngas in presence of ionic liquids [15]. Traditionally, the syngas is a mixture of carbon monoxide and hydrogen. The industrial use of carbon monoxide is not without its hazards, including toxicity risks which require special precautions and potential flammability when mixed with air. The entire process is mediated with catalysts which are usually derived from cobalt or rhodium. The new process patented by Evonik utilizes carbon dioxide instead of carbon monoxide in the syngas mixture. This process is more attractive because it can transform a notable greenhouse gas and contributor to global warming.

The Evonik process [16] utilizes novel supported ionic liquid phase (SILP) technology to immobilize the precious metal catalyst on a porous material (Figure 4.3). A homogeneous catalyst is dissolved in an ionic liquid thin film which is then applied to the surface of a porous, inert solid material by physisorption. The advantage here versus prior approaches is the need for less ionic liquid and reduction in mass transfer effects. Furthermore, the ionic liquid can have a positive effect on the kinetics of the process, favoring the water gas shift reaction for formation of CO from CO_2 due to excellent solubility of the latter in ionic liquid.

Recently, Evonik announced via press release that the SILP rhodium catalyst system can be used commercially by demonstrating a successful pilot scale run with considerable extended service life of 2,000 hours for the catalyst system. The ionic liquid is said to be based on the imidazolium cation with a binary amine anion [17].

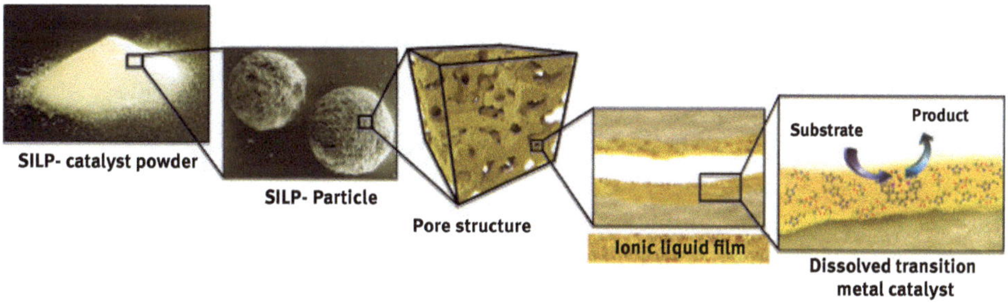

Figure 4.3: Hydroformylation reaction with SILP. Reprinted with permission from Evonik Industries AG.

4.3 Polymer processing

4.3.1 Cellulose and biomass

Cellulose is a very abundant biorenewable resource with on the order of hundreds of billions of tons of the natural polymer present globally [18]. Naturally, because of its abundance, people have developed methods for separating it from biomass and utilizing it for a host of purposes [19]. Cellulose has negligible solubility in water. Therefore several solvent systems have been developed for processing cellulose, including the viscose process [20] and the NMMO (Lyocell) process [21]. In the former process, the cellulose is chemically modified via treatment with NaOH, derivatization with carbon disulfide, and regeneration with acid solution. An obvious drawback is degradation of the cellulose backbone. The latter process utilizes N-methylmorpholine oxide (NMMO) as a non-derivatizing solvent, but suffers from the drawback of cost and the thermal instability of the solvent.

Interestingly, the use of ionic liquids in cellulose processing dates back to 1934 with a patent by Graenacher [22] in which he claims the use of molten quaternary ammonium salts to directly prepare solutions of cellulose. Fast forward to 2002 when Swatloski et al. found [23] (and patented) [24] that ionic liquid based on combinations of imidazolium cations and halide anions could directly dissolve cellulose with only mild heating or microwaving. Since then, numerous researchers have found a multitude of ionic liquid combinations and conditions that can dissolve cellulose [25]. Most notable are 1,3-dimethylimidazolium acetate and 1-butyl-3-methylimidazolium chloride which can provide clear, homogeneous cellulose solutions as high as 25 wt%. It is now well-established [26] that the anion of the ionic liquid favorably interacts via hydrogen bonding with the hydroxyl protons of cellulose, disrupting the strong intermolecular hydrogen bonds that exist between carbohydrate chains, thus favoring dissolution (Figure 4.4).

The ability to directly and easily dissolve cellulose opens up many opportunities to process the cellulose into different forms and materials like fibers and films. Because cellulose is water insoluble and the favored ionic liquids are water soluble,

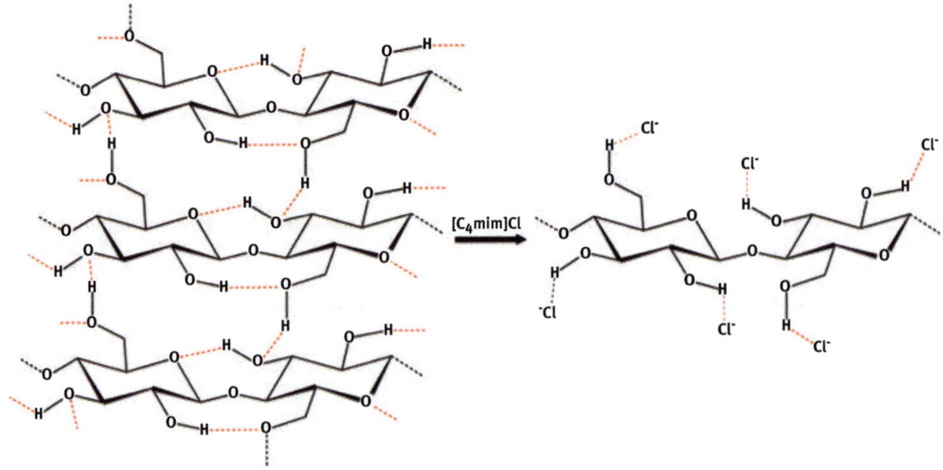

Figure 4.4: Dissolution of cellulose with ionic liquids.

it is quite easy to regenerate cellulose by treatment of the mixture with solvents like water and ethanol, washing away the salt and leaving behind cellulose in the process. In fact, this is so attractive that in 2006 BASF and the University of Alabama formalized a partnership for BASF to exclusively license the patents to dissolve, regenerate, and process cellulose using ionic liquids [27].

In addition to direct cellulose dissolution, ionic liquids have also been investigated for direct processing of lignocellulosic biomass to separate the lignin, hemicellulose, and cellulose components. In 2007 [28], it was first demonstrated that 1-butyl-3-methylimidazolium chloride is capable of processing untreated woods and the cellulose can be reconstituted with a variety of precipitating solvents virtually free of lignin and hemicellulose. Thereafter, many studies have investigated the extent to which biomass processing can be achieved [25].

4.4 Metal processing

4.4.1 Electroplating of aluminum

Ionic liquids are attractive for electrochemical applications, particularly metal plating and deposition, because of their broad liquid range, excellent stability, and notably their wide electrochemical window. “Electrochemical window” refers to the voltage range that can be applied across a material before the substance undergoes oxidation or reduction. The electrodeposition of aluminum in what could be called ionic liquids – N-alkyl pyridinium halides – was first reported as long ago as 1944 [29]. Due to the explosive growth of ionic liquids recently, the commercial applicability of such a method has once again been taken up. Ionic liquids are attractive for this purpose because the aqueous electrolyte solutions that are typically used in metal plating cannot be used for aluminum due to the electrochemical window of the solutions being too narrow (resulting in electrolysis). While water is adequate for metals like copper, silver, gold, chromium, and nickel, other metals like aluminum and titanium require something with a broader range. An alternative method has been developed but the process is difficult to implement [30]. BASF has demonstrated a laboratory scale process utilizing 1-ethyl-3-methylimidazolium chloride that is more robust and claims a number of advantages, including operating temperatures of between 60 and 100 °C, good water tolerance to 0.1 wt%, excellent quality, adhesion, and workability of the resultant coatings on a variety of base substrates, and low ionic liquid losses [31].

4.5 Separations

4.5.1 Nuclear industry

Decades of U.S. nuclear weapons development and nuclear power generation for commercial and residential purposes have resulted in immense quantities of highly radioactive nuclear waste. Due to policies in the U.S. that prevent recycling and reprocessing, this waste is stored on site at government lab locations like Hanford, Washington and at nuclear power plants scattered across the country. Because of the highly radioactive nature of the waste, countless discussions and debates have occurred about what to do with it. For example, the components that make it highly radioactive, i. e., the nuclides with short half lives like ^{90}Sr and ^{137}Cs, could be removed thus lowering the volume and subsequent storage requirements of this highly dangerous and hazardous portion. Furthermore, the fuel rods used in nuclear power plants still retain a significant fraction of fissile elements that could be recovered and recycled into new rods. Once again, because of the contentious nature of this highly political topic and concerns about proliferation, these approaches to recovery and treatment are nearly always discounted and short-term storage or long-term interment (like the now defunct Yucca Mountain) via glassification, for example, always win the day [32].

Nonetheless, options for nuclear waste treatment via chemical means are important to discuss and study because storage is often a highly contested issue (think "not in my backyard") and the current situation may evolve to make it a viable option. For example, France has a well-developed recycling program to recover useful fuel, dispose of low-volume waste, and reconstitute high value and energy-dense materials into new fuel rods [33]. Well-studied and established methods are in place, like PUREX, TRUEX, DIAMEX, for recycling and recovery, some of which were highly utilized during the Manhattan Project to isolate uranium and plutonium for weapons purposes [34]. These methods have a tremendous solvent requirement and produce copious waste, including nitric acid, tributyl phosphate, and paraffinic hydrocarbons, among others. These solvents also have use concerns, including radiolytic stability, flammability, and volatility.

Room temperature ionic liquids have always been of high interest for nuclear-type separations and processing [32]. The reasons are many: low volatility, non-flammability and combustibility, wide liquid range, thermal stability, radiolytic stability, [35] potential for recycling, the ability to make hydrophobic ionic liquids for biphasic separations, and the opportunity to make task-specific ionic liquid for sequestration. [36, 37], For performing separations, ionic liquids based on hexafluorophosphate (PF_6^-) and bis(trifluoromethanesulfonyl)imide (Tf_2N^-) anions have received attention because they form biphasic systems when contacted with water [38]. However, the PF_6^- anion has been shown to undergo hydrolysis to form HF [39]. Aqueous biphasic systems with hydrophilic ionic liquids are also possible in high salt environments via a salting-out process [40]. In both instances, these ionic liquids can be used as extracting phases when contacted with aqueous nuclear waste solutions containing metal ions of interest.

Dai et al. have shown that large distribution ratios are possible for fission products like $^{90}Sr^{2+}$ utilizing crown ethers in imidazolium-based PF_6^- and Tf_2N^- ionic liquids. The effects were as high as 4 orders of magnitude greater than with other extraction solvents like toluene [41]. Separation of actinides like Am^{3+}, Pu^{4+}, Th^{4+}, and $(UO_2)^{2+}$ in similar ionic liquids has also been demonstrated using extractants such as CMPO (octyl(phenyl)-N,N-diisobutylcarbamoylmethylphosphine oxide) yielding higher distribution ratios than with typical solvents like n-dodecane [42]. Technically speaking, the resolution or separation of trivalent lanthanides from actinides (TALSPEAK) is quite challenging; however, it becomes feasible with higher efficiencies and selectivities when ionic liquids are used versus conventional solvents [43]. Similar to the aluminum example discussed above, ionic liquids are also attractive for the electrodeposition of actinides and lanthanides because of the large voltage potential that is required but not attainable with many other solvents [44]. These are just a few of the numerous examples that one can find in the scientific literature for the use of ionic liquids which continue to offer an attractive outlook for separations and processing efforts relating to the nuclear fuel cycle.

4.6 Functional fluids

4.6.1 Gas compression

Ionic liquids exhibit many interesting properties, not least among them are (virtual) incompressibility and unmeasurable vapor pressure. Taking advantage of these unique attributes, Linde has developed, patented [45], and commercially demonstrated an ionic compressor for next generation hydrogen fueling technologies. In the compressor, the ionic liquid replaces the traditional metal piston [46]. Gas is compressed in the cylinder by the up-and-down motion of the specially designed ionic liquid piston (similar to a traditional piston) in which hydrogen gas is negligibly soluble (as compared to other molecular solvents) [47]. Furthermore, because ionic liquids exhibit no vapor pressure, there is no mixing of the ionic liquid solvent of the piston in the hydrogen gas. For these reasons, including the wide liquid range of the ionic liquid, the compressor can be adjusted to meet a range of requirements and is safe, reliable, and economical for fueling vehicles at 35 or 70 MPa.

Linde cites a number of highlights and benefits for the ionic compressor: no gas contamination, no volatility, no combustibility, environmentally safe, good thermal capacity, good lubricating qualities, no corrosion, and the elimination of seals and bearings. The Ionic Compressor 90 MPa – IC90 has an approved design, is easy to operate, and is in service in various locations around the world. It uses a 5-stage compression system that meets the SAE J 2601 fueling standard and allows for continuous, fast, and high-performance fueling of hydrogen vehicles at low operating cost, low energy consumption, near 100 % energy conversion efficiency, and a very small number of moving parts. Linde has positioned itself as leader in hydrogen fueling technology utilizing ionic liquids to meet the global needs of the evolving transportation sector.

4.6.2 Gas capture and storage

As described in Section 6.1, several gases including hydrogen have very low solubility in ionic liquids as compared to molecular solvents. However, other gases like carbon dioxide exhibit rather large solubility in ionic liquids [48]. Blanchard et al. were the first to show that imidazolium-based ionic liquids exhibited high solubilization capacity for carbon dioxide; furthermore, the carbon dioxide product could be recovered with no contamination of the insoluble ionic liquid [49]. The high inherent solubility has several explanations. Early ATR-IR data [50] suggested that a favorable Lewis acid-base interaction exists between CO_2 and the anion, such as PF_6^- and BF_4^- with the 1-butyl-3-methylimidazolium cation. Free volume arguments are also often invoked in experimental [51] and simulation [52] interpretations and likely play a significant role in CO_2 gas solubility. Traditional ionic liquids as sorbents for CO_2 are attractive because of their tunable nature to maximize the favorable interactions for the solubility of the gas using regular solution theory [53].

Ionic liquids are appealing solvents (compared to alkanolamines) for carbon dioxide capture in flue-gas from coal fired power plants because of their environmentally friendly profiles, particularly low volatility, and potential to be recycled/recovered. However, even though solubility of CO_2 in ionic liquids is relatively high, absorption capacity at low post-combustion partial pressures is lower than 5 mol% even with the most ideal ionic liquid. Therefore, researchers have modified ionic liquids to be reactive with respect to carbon dioxide and increase the molar

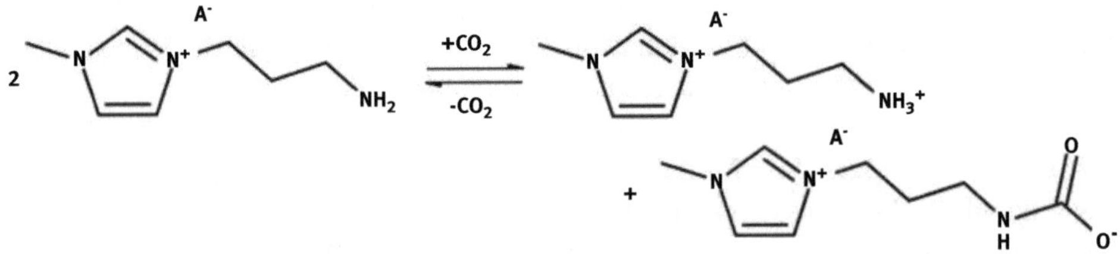

Figure 4.5: Amine-functionalized TSIL for CO_2 capture.

absorption capacity of these materials. Bates et al. [36], as described previously, were the first to make an amine-functionalized task specific ionic liquids (TSILs) and demonstrate that binding occurs in a molar stoichiometry of 2:1 ionic liquid:CO_2 (Figure 4.5). In this case, the amine was appended to the cation and the mechanism proceeded via formation of a thermally reversible carbamate/ammonium pathway. However, the requirement of 2 moles of ionic liquid cation for every 1 mole of CO_2 is inefficient (relative to a 1:1 stoichiometry, for example) and the reaction results in a highly viscous capture medium, making processing and recovery nearly impossible. Gutowski and Maginn [54] demonstrated via molecular dynamics simulations that this high viscosity is due to the formation or a pervasive hydrogen bonding network between the carbamate and ammonium moieties.

More recently, progress has been made utilizing a new approach that incorporates the CO_2 binding functionality of the ionic liquid into the anion. In a combined study, it was predicted from simulation and demonstrated from experiment that aprotic heterocyclic anions, like 2-cyanopyrrolide, can overcome the deficiencies of amine-functionalized cations when paired with tetrabutylphosphonium (simulation) or trihexyl(tetradecyl)phosphonium (experiment) cations [55]. It was confirmed from *in situ* infrared spectroscopy that a carbamate is formed with a 1:1 anion:CO_2 stoichiometry, making it a much more efficient complexing TSIL. Also, the measured viscosity of the reacted and unreacted TSIL was identical, indicating that the reaction mechanism is indeed different versus the amine cation. The authors suggested that anions such as imidazolide and pyrazolide will also show similar behavior and was demonstrated with the synthesis of the latter ionic liquid. Despite no differences in viscosity upon complexation, the viscosity of the neat ionic liquid was quite high, perhaps precluding it from commercial usage. As provided in these examples, considerable progress has been made in this area and it is encouraging that tuning of the ionic liquids to achieve suitable viscosities and binding enthalpies is in reach. Nevertheless, the scale of the carbon dioxide capture problem in coal fired power plants is immense and it remains to be seen if ionic liquids and TSILs can play a role in mitigating this ever-growing problem.

Ionic liquids also offer opportunities for safe storage of other hazardous and reactive gases, including boron (III) fluoride, phosphine, and arsine. On this front, air products have developed a technology utilizing Lewis base ionic liquids (BF_4^-) for Lewis acid gases (BF_3) and vice versa (Cu_2Cl_3 and Cu_2Br_3 ionic liquids for phosphine and arsine). This process, called Gasguard Sub-Atmospheric System, Complexed Gas Technology (CGT), reduces the risk associated with other methods by delivering the gases in a safe, effective, and more easily handled way [56].

4.7 Other (potential) applications

While it is fair to say that over the years ionic liquids have been considered for a number of diverse applications, in some cases the extent of their implementation, or realization of their usage entirely, is yet to be determined. In the opinion of the author, ionic liquids have evolved from a class of materials with broad hopes and aspirations to a subset of specialty chemicals with niche utility. In this final section, some of these niche usages will briefly be described.

4.7.1 Dye-sensitized solar cells

Dye-sensitized solar cells (DSSCs) are devices used to generate electricity from solar energy (visible light) and offer the potential to be significantly cheaper and easier to manufacture than traditional semiconductor based devices. DCCSs of the Graetzel type [57] utilize a molecular dye that traps solar energy passing through the transparent electrode and transfers its excited electrons to nanoparticulate TiO_2. The space between the electrodes contains a redox couple (based on iodide) in a liquid electrolyte that transports charge passing from the counter electrode back to the dye (Figure 4.6). To this end, ionic liquids have been investigated as liquid electrolytes. When subjected to repeated cycles of thermal heating, volatility of the solvent electrolyte can be a major

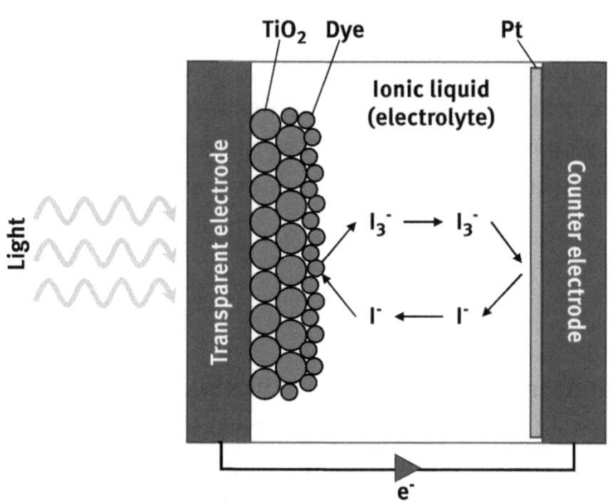

Figure 4.6: Dye-sensitized solar cell utilizing ionic liquid electrolyte.

factor in developing a stable DSSC. Non-volatile solvents like ionic liquids therefore offer the promise of stable devices with long-term performance [58]. In 2007, G24 Innovations and BASF announced a partnership to develop ionic liquids as novel electrolytes in DSSCs with the goal of improving the performance and efficiency of G24i's solar cells [59]. As of 2017, little new information is available.

4.7.2 Batteries

Similar to DSSCs, ionic liquids also offer promise as electrolytes in other electrochemical devices like next generation lithium ion batteries. Lithium ion batteries are ubiquitous and power all of our mobile devices from laptops to cell phones to electric cars. However, it is generally agreed that improvements are continually needed to meet the ever-growing demand of increased cycling rate and battery lifetime. Ionic liquids are attractive primarily because of their large electrochemical windows, thus being suitable for energy dense materials and expansion beyond just lithium ion batteries to include metal-air batteries. Furthermore, their low vapor pressures and non-flammable nature makes them inherently safer and stable materials for this application. Despite the promise, commercial implantation of ionic liquids in modern battery technology has yet to be realized [58].

4.7.3 Antistatics

Ionic liquids have found use as antistatic agents in resins, coatings, and cleaning agents. Antistatics are important to prevent static electricity from building up on surfaces. The 3M company has commercialized ionic liquids as high purity antistatic additives compatible with a variety of high performance polymer systems, including urethanes, acrylates, thermoplastic, and thermoset resins. These materials provide optical clarity, minimal metal and halogen ion content for low corrosivity, stable performance over a wide humidity range, static dissipation in a wide variety of polymers, negligible vapor pressure, and hydrophobic properties for water/washing resistance [60]. Similarly, BASF has marketed ionic liquids as antistatic additives for coatings, providing materials that are transparent, liquids, non-toxic, and available as conductive particles [61].

4.7.4 Consumer packaged goods

The use or potential use of ionic liquids in consumer products like personal care and household cleaning compositions is seldom considered in reviews. Ionic liquids are promising materials for this type of end use for many reasons, including the ability to be tuned or functionalized via cation and/or anion modification for particular benefits, their ability to be media for the delivery or solvation of actives, and their traditionally "green" attributes like no vapor pressure. The Procter & Gamble company has arguably been the most active in this area, as evidenced by numerous patents and patent applications protecting ionic liquid-based inventions.

One of the first patent applications describes ionic liquid-based products and method of using the same [62]. Here, the ionic liquid "cocktails" containing three or more different charged ionic liquid components are mentioned as being useful in surface treating, air treating, and cleaning. P&G also has noted that ionic liquids can possess amphiphilic properties and therefore can be derived to behave like functionalized anionic surfactants [63]. Surfactants are the principal ingredients in cleaning detergents for fabric, hair, and other surfaces. Furthermore, certain ingredients like oxygen bleach are not stable in liquid detergent environments. Ionic liquids can be designed to contain a peracid anion as a novel and unique way to stabilize and deliver bleach in detergents [64]. In addition to surfactants for cleaning, P&G has noted that ionic liquids are useful for delivery of ion actives capable of delivering benefits, for example, a fabric treating benefit, a surface treating benefit, and/or an air treating benefit, to a target substrate [65]. Another mechanism for "actives" is also possible with ionic liquids where the active that is protected or stabilized by an ionic liquid is releasable from the system to provide the benefit. For example, the ionic liquid may be physically or chemically associated with the benefit agent [66]. Finally, because ionic liquids are tunable in many ways, including miscibility with water and with each other, multiphase cleaning products are accessible where each individual phase is capable of delivering or providing a benefit [67]. While it is currently unknown if any of the aforementioned uses of ionic liquids has reached the commercialization stage, it should be obvious to the reader that the benefits are many in the consumer good arena and it seems inevitable that such products will reach the market.

4.8 Outlook

In the early 2000s, ionic liquids as a class of materials and a field of scientific inquiry launched into our collective field of view promising to be "infinitely" tunable with the inevitability of solving a host of the worlds' problems – all with a "green" nameplate. However, as time has evolved and the hype has died down, it has become undeniably obvious that ionic liquids are no longer the miracle solvents of the future and the solution for all our industrial problems. If this were the case, industrial implementation would be widespread and commercialization toward improving the bottom line would be a no-brainer. Instead, ionic liquids have now reached a point where academic evaluations are becoming more focused and industrial consideration more niche. Rather than being "materials for all", ionic liquids have now become materials "where it makes sense", i. e. where leveraging the numerous interesting and useful properties that make them unique is most practical. As shown in this chapter, these practical industrial uses and applications are many. In some cases, implementation has reached realization and industry is moving full steam ahead. In other instances, ionic liquids have still not attained their time and place, despite the community backing and the scientific evidence that lends support. In the latter cases, the hype is not enough. However, it seems inevitable in many of these examples that ionic liquids will take their position in the sun. Arguably, we have yet to scratch the surface, not with diversity of consideration, but rather with industrial implementation. The possibilities continue to be endless. Plechkova and Seddon in a similar review from 2008 [68], predicted that ionic liquids will find commercial implementation in nearly 26 applications covering six different broad categories. Nearly 10 years later, implementation in many of these areas is yet to be realized, but perhaps we have only just begun to make inroads.

Notes: The author would like to thank Dr. Martin Johannes Schneider for providing the SILP graphic in Figure 3.

References

[1] Gabriel S, Weiner J. Ber Dtsch Chem Ges. 1888;21:2669–79.
[2] Walden P. Bull Russian Acad Sci. 1914;8:405–22.
[3] Haynes WM, ed. CRC handbook of chemistry and physics. 94th ed. Boca Raton: FL: CRC Press LLC; 2013-2014:4–89.
[4] Abbott Andrew P., Capper Glen, Davies David L., Rasheed Raymond K., Tambyrajah Vasuki. Novel solvent properties of choline chloride/urea mixturesElectronic supplementary information (ESI) available: spectroscopic data. See http://www.rsc.org/suppdata/cc/b2/b210714g/. Chemical Communications. 2002 11 26;(1):70–71. DOI: 10.1039/B210714G
[5] Welton T. Chem Rev. 1999;99:2071–83.
[6] Wilkes John S., Zaworotko Michael J. Air and water stable 1-ethyl-3-methylimidazolium based ionic liquids. Journal of the Chemical Society, Chemical Communications. 1992;(13):965–965. DOI: 10.1039/C39920000965
[7] Gale R. J., Osteryoung R. A. Potentiometric investigation of dialuminum heptachloride formation in aluminum chloride-1-butylpyridinium chloride mixtures. Inorganic Chemistry. 1979 6; 18(6):1603–1605. DOI: 10.1021/ic50196a044
[8] Wassersheid P, Welton T, eds. Ionic liquids in synthesis, 2nd ed., vol. 1. Wiley-VCH, 2008.
[9] Earle MJ, Seddon KR. Pure Appl Chem. 2000;72:1391–98.
[10] Maase M, Massone K, Halbritter K, Noe R, Bartsch M, Siegel W, et al. U.S. Patent 7,351,339 B2, 2003.
[11] Maase M. Industrial applications of Ionic liquids. In: Ionic liquids in synthesis, 2nd ed., vol. 1. Weinheim, Germany: Wiley-VCH Verlag GmbH & Co. KGaA, 2008:663–87.
[12] Hell K, Weyershausen B, Hesses U U.S. Patent Application US 20040014925 A1, 2003.
[13] Gelbach TJ, Zhao D, Castillo NC, Laurenczy G, Weyerhausen B, Dyson PJ. J Am Chem Soc. 2006;128:9773–80.
[14] Franke R, Fridag D, Hamers B, Zanthoff H-W, Blug M, Strautmann J, et al. U.S. Patent Application US 20130178657 A1, 2011.
[15] Haumann M, Riisager A. Chem Rev. 2008;108:1474–97.

[16] Schönweiz A, Franke R. Supported ionic liquid phase (SILP) materials in hydroformylation catalysis. In: Fehrmann R, Riisager A, Haumann M, eds. Supported ionic liquids: fundamentals and applications. Weinheim, Germany: Wiley-VCH Verlag GmbH & Co. KGaA; 2014.
[17] Evonik demonstrates for the first time that SILP catalyst systems can be used commercially. *Evonik Industries* Press Release, June 10, 2015, Essen, Germany,
[18] Klemm D, Heublein B, Fink H-P, Bohn A. Angew Chem Int Ed. 2005;44:3358–93.
[19] Edgar KJ, Buchanan CM, Debenham JS, Rundquist PA, Seiler BD, Shelton MC, et al. Prog Polym Sci. 2001;26:1605–88.
[20] De Wyss G. Ind Eng Chem. 1925;17:1043–45.
[21] Dawson T. Coloration Technol. 2011;128:1–8.
[22] Graenacher C U.S. Patent 1,943,176, 1931.
[23] Swatloski RP, Spear SK, Holbrey JD, Rogers RD. J Am Chem Soc. 2002;124:4974–75.
[24] Swatloski RP, Holbrey JD, Rogers RD U.S. Patent 6,824,599 B2, 2002.
[25] Wang H, Gurau G, Rogers RD. Chem Soc Rev. 2012;41:1519–37.
[26] Remsing Richard C., Swatloski Richard P., Rogers Robin D., Moyna Guillermo. Mechanism of cellulose dissolution in the ionic liquid 1-n-butyl-3-methylimidazolium chloride: a 13C and 35/37Cl NMR relaxation study on model systems. Chemical Communications. 2006;(12):1271–1271. DOI: 10.1039/b600586c
[27] BASF, UA formalize ionic liquids partnership. UA News Press Release, March 8, 2006, Tuscaloosa, AL.
[28] Fort DA, Remsing RC, Swatloski RP, Moyna P, Moyna G, Rogers RD. Green Chem. 2007;9:63–69.
[29] Hurley FH U.S. Patent 2,446,331, 1944.
[30] Ziegler K, Lehmkuhl H. Chemie. 1956;283:414–25.
[31] O'Meara M, Alemany A, Maase M, Vagt U, Malkowsky I. Deposition of aluminum using ionic liquids. *Materials Today*, 10 August 2009. http://www.materialstoday.com/metal-finishing/features/deposition-of-aluminum-using-ionic-liquids/
[32] Sun X, Luo H, Dai S. Chem Rev. 2012;112:2100–28.
[33] http://www.euronuclear.org/info/encyclopedia/r/reprocessing-plants-ww.htm. Accessed February 13, 2018.
[34] Lumetta GJ, Nash KL, Clark SB, Friese JI. Separations for the nuclear fuel cycle in the 21st century. Washington, DC: American Chemical Society; 2006.
[35] Allen D, Baston G, Bradley AE, Gorman T, Haile A, Hamblett I, et al. Green Chem. 2002;4: 152–158.
[36] Bates ED, Mayton RD, Ntai I, Davis J. H. Jr. J Am Chem Soc. 2002;124:926–27.
[37] Visser Ann E., Swatloski Richard P., Reichert W. Matthew, Davis Jr. James H., Rogers Robin D., Mayton Rebecca, Sheff Sean, Wierzbicki Andrzej, et al. Task-specific ionic liquids for the extraction of metal ions from aqueous solutions. Chemical Communications. 2001;(1):135–136. DOI: 10.1039/B008041L
[38] Huddleston Jonathan G., Willauer Heather D., Swatloski Richard P., Visser Ann E., Rogers Robin D. Room temperature ionic liquids as novel media for 'clean' liquid–liquid extraction. Chem. Commun. 1998;(16):1765–1766. DOI: 10.1039/A803999B
[39] Swatloski RP, Holbrey JD, Rogers RD. Green Chem. 2005;5:361–363.
[40] Gutowski KE, Broker GA, Willauer HD, Huddleston JG, Swatloski RP, Holbrey JD, et al. J Am Chem Soc. 2003;125:6632–33.
[41] Dai S, Ju YH, Barnes CE. Dalton Trans. 1999;8:1201–02.
[42] Visser AE, Rogers RD. J Solid State Chem. 2003;171:109–13.
[43] Sun X, Bell JR, Luo H, Dai S. Dalton Trans. 2011;40:8019–23.
[44] Binnemans K. Chem Rev. 2007;107:2592–614.
[45] Adler R, Pfandi M, Siebert G U.S. Patent 9,109,587 B2, 2011.

[46] Linde company website. "Hydrogen Technologies: The Ionic Compressor 90 MPa – IC90" brochure.
[47] Dyson Paul J., Laurenczy Gábor, André Ohlin C., Vallance James, Welton Thomas. Determination of hydrogen concentration in ionic liquids and the effect (or lack of) on rates of hydrogenation. Chem. Commun. 2003;(19):2418–2419. DOI: 10.1039/B308309H
[48] Ramdin M, De Loos TW, Vlugt TJH. Ind End Chem Res. 2012;51:8149–77.
[49] Blanchard LA, Hancu D, Beckman EJ, Brennecke JF. Nature. 1999;399:28–29.
[50] Kazarian Sergei G., Briscoe Brian J., Welton Thomas. Combining ionic liquids and supercritical fluids: in situ ATR-IR study of CO2 dissolved in two ionic liquids at high pressures. Chemical Communications. 2000;(20):2047–2048. DOI: 10.1039/B005514J
[51] Blanchard LA, Gu Z, Brennecke JF. J Phys Chem B. 2001;105:2437–44.
[52] Cadena C, Anthony JL, Shah JK, Morrow TI, Brennecke JF, Maginn EJ. J Am Chem Soc. 2004;126:5300–08.
[53] Shi W, Maginn EJ. J Phys Chem B. 2008;112:16710–1670.
[54] Gutowski KE, Maginn EJ. J Am Chem Soc. 2008;130:14690–704.
[55] Gurkan BE, De La Fuente J, Mindrup EM, Ficke LE, Goodrich BF, Price EA, et al. J Am Chem Soc. 2010;132:2116–17.
[56] Freemantle M. Chapter 14: applications. In: An introduction to ionic liquids. RSC Publishing; 2010. ISBN 978-1-84755-161-0.
[57] O'Regan B, Grätzel M. Nature. 1991;353:737–40.
[58] MacFarlane DR, Tachikawa N, Forsyth M, Pringle JM, Howlett PC, Elliott GD, et al. Energy Environ Sci. 2014;7:232–50.
[59] BASF and G24i develop solar cells. Chem Eng News. 2077;85:25.
[60] 3M™ company website. Antistatic Additives for Electronics.
[61] BASF company website. "Basionics: high performance antistatic additives" brochure. http://www.intermediates.basf.com/chemicals/ionic-liquids/. Accessed February 13, 2018.
[62] Price K, Hartshorn R, Rohrbaugh R, Scheper W, Showell M, Baker K, et al. U.S. Patent Application 20040077519A1, 2003.
[63] Hecht SE, Cron SL, Scheibel JJ, Miracle GS, Seddon KR, Earle M, et al. U.S. Patent 7,737,102, 2005.
[64] Hecht SE, Miracle GS, Cron SL, Showell MS U.S. Patent 7,786,065, 2006.
[65] Kenneally CJ, Hecht SE, Cron SL U. S. Patent 7,737,106 B2, 2006.
[66] Price KN, Hecht SE, Jordan GT, IV, Kenneally CJ, Aouad YG, Showell MS, et al. U.S. Patent 7,939,485 B2, 2005.
[67] Hecht SE, Price KN, Berger PS, Foley PR, Hutton HD, III, Showell MS, et al. U.S. Patent 7,928,053, 2009.
[68] Plechkova NV, Seddon KR. Chem Soc Rev. 2008;37:123–50.

Philip G. Sliva

5 Greener chemistry opportunities in consumer products rinse added fabric softener

Abstract: Opportunities for greener chemistry seem endless. Consumer products, which represent very large users of chemical materials, are no exception. Even products such as fabric softeners that have existed for decades need continual enhancement. Many improvements have been implemented in the chemistry used in these products; however, more is warranted. Key challenges include formula stability, raw material sourcing based on animal fats, and inability to produce concentrated, lower dose products.

Keywords: Green Chemistry, Fabric Softener, Viscosity, Triethanolamine, Ester Quat, Esterquat Stability, Biodegradation

5.1 Background

Liquid rinse added fabric softeners have existed since the mid 1950s. They were created to meet a consumer need caused by the improvements of the cleaning agents used for laundering clothing. Prior to the middle of the twentieth century, tallow soaps were the primary cleaning agent used for laundering. Tallow soaps tended to deposit on fabrics during the laundry process and gave the dried fabrics a feeling of softness. These were replaced with synthetic-based detergents in the late 1940s. This conversion to synthetic detergents offered better cleaning and whiteness, but because fabrics were no longer "lubricated" with the tallow soap deposits, they tended to feel more harsh to the wearer when dried.

Consumer product goods companies recognized this new need and responded by developing a new product, rinse added fabric softeners. Globally this product category need was confirmed and has expanded into a large business. Euromonitor International [1] estimated sales at over $8 billion in 2010 with expectations of rapid increases as the consumer class in China continues to expand.

These products offer a number of benefits to consumers with the most obvious being improved fabric feel. In addition, they reduce the buildup of static electricity [2] on fabrics. Static electricity is primarily of concern in low humidity climates and most especially in markets that make use of automatic tumble dryers (note that automatic

This article has previously been published in the journal Physical Sciences Reviews. Please cite as: Sliva, P. Greener Chemistry Opportunities in Consumer Products Rinse Added Fabric Softener. *Physical Sciences Reviews* [Online] **2018**, *3*. DOI: 10.1515/psr-2017-0192

https://doi.org/10.1515/9783110562781-005

clothes driers are primarily limited to North America as the rest of the world typically air dries laundry which explains why the dryer sheet market is limited to $1 billion). Other fabric softener benefits include the extension of fabric life [3] and, when appropriately fragranced, the addition of a pleasant scent to laundry.

Formulating successful consumer products can be very challenging. The product must deliver the performance a consumer expects across a relatively long storage period where the product may be exposed to stressful conditions (very hot to very cold temperatures). As companies strive to improve the sustainability or "greenness" of consumer products a consistent trend has been to minimize product dose. This reduces total package needs and has added supply chain benefits as transportation impact is reduced. This goal has proven particularly challenging for fabric softener formulators.

5.2 Chemistry

There are many ancillary components used in fabric softeners; however, nearly all commercially sold fabric softeners are primarily based on cationic surfactants that are quaternary ammonium compounds (quats). For this discussion, the focus will be limited to these "quats".

Those familiar with surfactants will know these molecules are composed of a hydrophilic or water loving portion and a hydrophobic or water hating portion. The hydrophobes of these cationic compounds are normally based on the bio-renewable feed stocks of animal oils or seed oils. These are often hydrogenated to increase saturation to some degree to reduce propensity to oxidize over time and to improve their softening performance. As a general rule, increasing the degree of saturation will improve the perceived feel of the softened fabric. The resulting cationic surfactants are cost effective and are highly efficient at exhausting from the rinse water onto fabrics even when formulated at low concentrations. The cationic nature of the molecules delivers especially effective deposition on negatively charged cellulosic (cotton) fibers.

There have been numerous types of active softeners, but these can be separated by major classes. Table 5.1 provides a summary of the materials and a rough estimate of performance of the key attributes of each type.

Table 5.1: Relative performance comparison.

Attribute:	Dialkyl dimethyl quat	Imidazolinium	Amido Amine Quat	Ester Quat
Softness	4	3	3	3
Static control	3	2	2	4
Concentration	1	2	4	4
Biodegradability	1	2	2	4
Shelf life	4	4	2	2

4 = Best Performance.

5.2.1 Dialkyl dimethyl quat

This is one of the earliest and most effective classes of commercial softener actives. It has superior performance for softening and controlling static. The limitations are an inability to formulate to high concentrations (unstable at room temperature at concentrations above ~15 % active level) and they are not considered to be biodegradable [4]. Fabric softeners formulated with dialkyl dimethyl quats are very unstable at low temperatures and will not recover from a freeze cycle. At high temperatures, they normally will phase separate. A commonly used example of this class is dihydrogenated tallow dimethyl ammonium chloride shown in Figure 5.1. It has steadily declined in use for fabric softeners, but has been used in some markets in Southeast Asia.

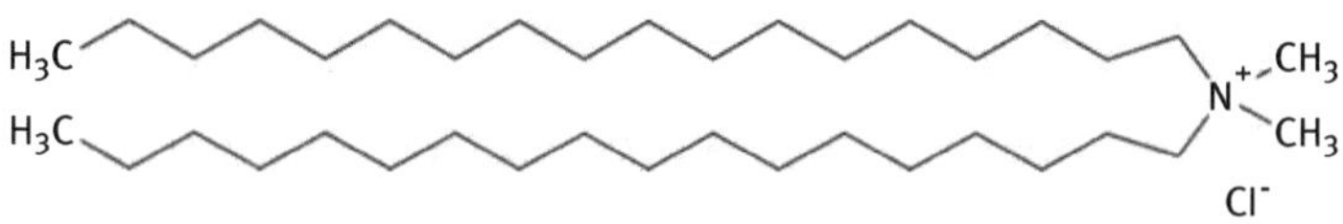

Figure 5.1: Dihydrogenated tallow dimethyl ammonium chloride.

5.2.2 Imidazolinium quat

Imidazolinium quats give reasonable performance, but are also limited in the ability to formulate to high concentrations. A depiction of a representative molecule is shown in Figure 5.2, where R represents an alkyl chain based on tallow. Because of this they were sometimes used in combination with DHTDMAC in some formulations. They are also not considered to be very biodegradable. Primarily due to their lack of biodegradability and their higher costs, they are not used in many formulations today.

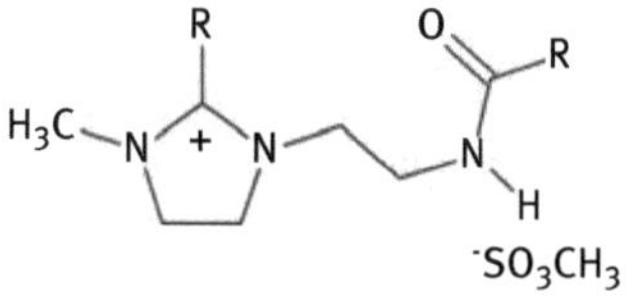

Figure 5.2: Imidazolinium quaternary ammonium methylsulfate.

5.2.3 Amidoamine quat

Most often this is a diamidoamine based on diethylenetriamine that is then ethoxylated and finally quaternized. Figure 5.3 is a representation of an ethoxylated diamidoamine that is quaternized with dimethylsulfate and R represents an alkyl chain based on tallow. This type of softener can be formulated to a very high concentration (over 20 % active). It has the disadvantage of poor low temperature stability and a

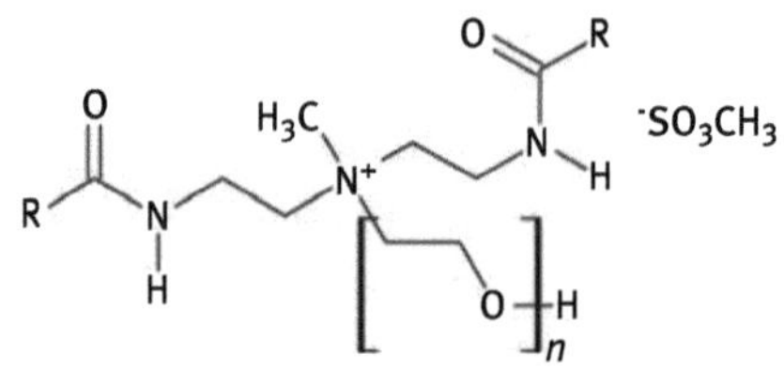

Figure 5.3: Diamidoamine ethoxylated quaternary ammonium methylsulfate.

tendency to increase in viscosity over time. It also is not considered to be very biodegradable. It too has steadily decreased in formulations over time.

5.2.4 Ester quat

Ester quats used in softener are primarily diesters of an alkanolamine that is then quaternized. Figure 5.4 shows a representation of a diester quat formed from triethanolamine (TEA) and tallow fatty acid, and quaternized and with dimethylsulfate. This type of active has become the primary fabric softener around the world because it is readily biodegradable. At the beginning of the twenty-first century, regulations accelerated its rise in use.

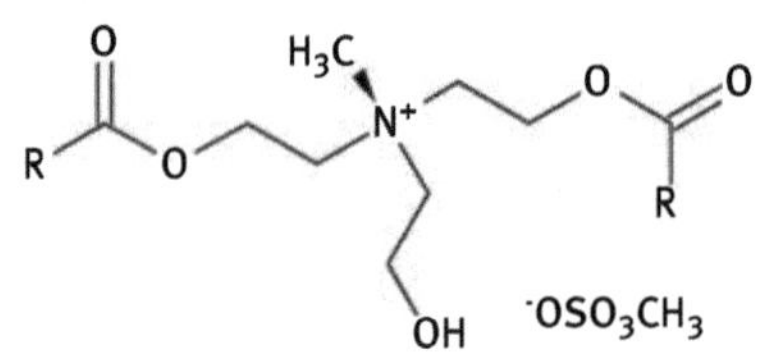

Figure 5.4: TEA esterquat.

As of October 2005, cationic surfactants were required by the European Union [5] to meet the minimum of 60 % mineralization in 28 days. The Organization for Economic Cooperation and Development (OECD) is leading international efforts to standardize biodegradation test methods. The OECD has published six tests on ready biodegradability. In general, a solution, or a suspension of a test substance in a mineral medium is inoculated and incubated under aerobic conditions. Biodegradation is determined by the amount of dissolved organic carbon in OECD 301 A and OECD 301E, the production of carbon dioxide in OECD 301 B, or the consumption of oxygen in OECD 301 C, OECD 301 D and OECD 301 F. The test will last 28 days; however, it can be ended sooner if the biodegradation ends sooner. Because ester quats meet this criterion, most manufacturers have moved to this technology and it has steadily been adopted around the world. A graphical representation of biodegradation is in Figure 5.5 (Andras Nagy, PhD – Evonik Industries) showing the relative difference between DHTDMAC (noted as

Dialkyl-dimethyl) and Esterquat (noted as Ester). Presented are the results of dissolved organic carbon via OECD 301 E at 5 days and oxygen consumption via OECD 301 C at 30 days.

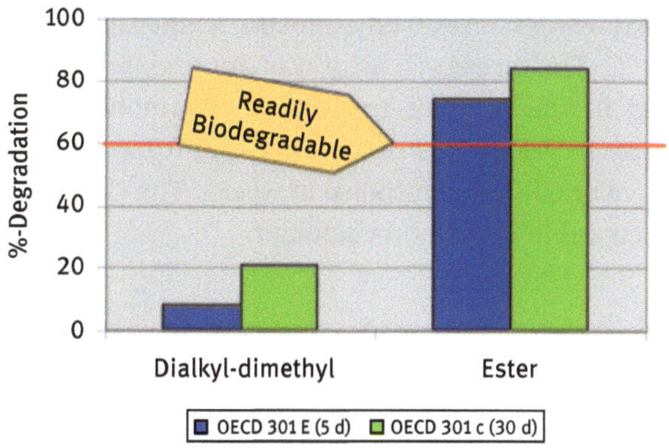

Figure 5.5: Biodegradation of quaternaries.

5.3 Formulation

Rinse added fabric softeners have always been produced essentially the same way regardless of how the cationic surfactants have evolved. With very few exceptions, rinse added fabric softeners are aqueous dispersions. As such they are also inherently unstable.

The cationic surfactants used as fabric softener actives are virtually water insoluble and nearly all of the highest performers, in concentrated form, are physically solid at temperatures below 120°F (50 °C). Even when melted the material does not flow well, requiring a solvent to be added (isopropanol, propylene glycol, or ethanol) to help in processing. As an example, most TEQ Esterquats are 85–90 % "active matter" with 10–15 % solvent.

To formulate a finished product, the highly concentrated softener active is heated until melted and slowly added to hot water with a high level of agitation to form tiny droplets. The sizes of the droplets are controlled by agitation speed. Once formed, the droplets are further reduced in size by adding a divalent salt. For high active formulas, this is repeated until the desired active concentration is achieved. After cooling and the droplets have hardened, colorant and fragrance can be added. In effect, the end product is finely dispersed solids in water.

In order to deliver rapid dispersion when added by the consumer to the laundry rinse water, some consumer product formulation chemists opt to produce fabric softeners with very low viscosity. This allows for a more uniform distribution of softener on fabrics. High viscosity products normally disperse slowly and need to be very low in actives to minimize localized spotting on fabric.

Unfortunately, the ester function in this softener, which allows it to be biodegradable and therefore a greener choice, makes it hydrolytically unstable. For the formulated fabric softener, this ultimately results in large viscosity increases, especially in high active formulas. To illustrate, samples of a market product from multiple production dates were stored in a warehouse without any temperature control. The age of the samples ranged from a few weeks to 900 days and the temperature over that period varied from 16.5 to 36.4 °C. The target viscosity was approximately 100 centipoise. Little change was noted in the samples until nearly 20 months had elapsed. As seen in Figure 5.6, the rise in viscosity is dramatic, rising from near the target 100 centipoise to near solid at over 60,000 centipoise in weeks. The change is so severe it makes the product unacceptable to the end customer.

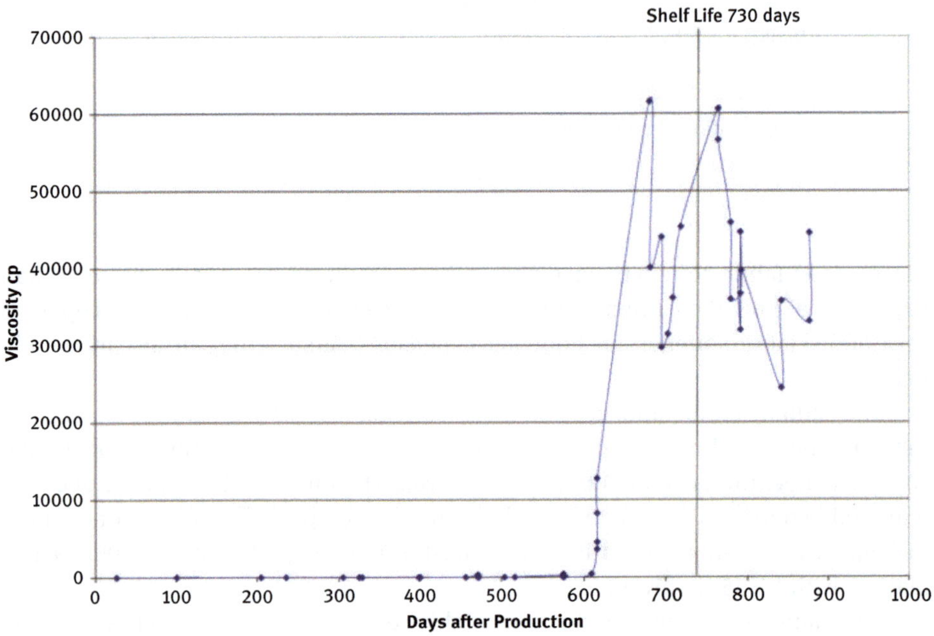

Figure 5.6: Fabric softener viscosity versus age in days.

These same fabric softener samples, which were produced starting in August 2006 through December 2008, were then analyzed for identity of the quaternary species. The analysis consisted of Atmospheric Pressure Chemical Ionization (ApCI) Mass Spectrometry (MS). A commercial esterquat (Stepantex VT-90 from the Stepan Company) was used to calibrate the method. This analysis showed the viscosity increase is related to the hydrolysis of the esterquat.

Figure 5.7 (courtesy of Stephen R. Missler, PhD – Amway Principal Research Scientist) is a comparison of the relative ester quat distribution with time and viscosity.

Note that the sudden increase in viscosity (orange bars) corresponds with declining diester quat (yellow bars) and increasing unesterified quat (blue bars) formation. This occurs at the May 2007 sample which was analyzed approximately 600 days after production. The initial increase in diester quat formation is due to a hydrolysis of triester quats (not shown). As triester and diester quats are hydrolyzed, an increase is observed in monoester quats, followed by a decline as hydrolysis continues.

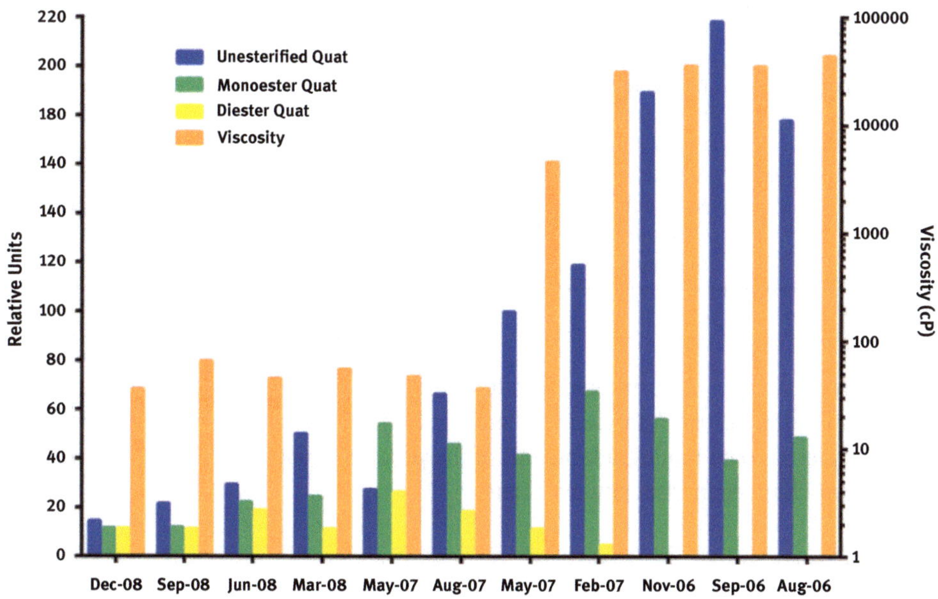

Figure 5.7: Comparison of ester quat distribution in fabric softener with time and viscosity.

For esterquat formulations, hydrolytic stability of the ester bonds has always been a concern. Once dispersed into the aqueous phase of the finished product, ample water is available to hydrolyze the esters. Theories on fatty acid release eliminating the electrostatic particle repulsion have been proposed [6]. However, it is just as likely that the reduced melting point of the resulting monoester or unesterified quat causes coalescence of individual particles into a viscous mass. Chemistry, concentration, time, and temperature are the factors that will affect stability. To avoid this, most manufacturers rely on lower active formulas and relatively short shelf lives.

This type of issue occurs in markets around the globe and is accelerated by temperature. As an example, a comparison of products sold in Malaysia and Thailand were evaluated for viscosity after storage at different temperatures. There is a variety of level of active expressed as solids (analyzed by evaporation of volatile materials) from as little as 3.0 % to as high as 22 %. Table 5.2 summarizes the resulting viscosity of each fabric softener after 3 months at varying temperatures as measured by Brookfield Viscometer and reported in centipoise.

Table 5.2: Viscosity after 3 months storage.

Manufacturer	Brand Name	Solids (%)	40°FcPs	70°F	100°F	120°F
Unilever	Comfort Happy (blue) [7]	3.0	11800	2347	950	Gel
IP Manufacturing LTD	Hygiene Pink Sweet [8]	3.9	127	120	270	Separated
United Detergent Industries	Petals Super Concentrated Brightening [9]	9.6	110	137	250	Gel
Procter & Gamble	Downy Nature Fresh [10]	10.2	10678	4300	67486	Gel
Unilever	Comfort Ultra [11]	10.4	14262	649	636	2449
Colgate-Palmolive	Softlan Ultra [12]	15.1	5753	7438	Gel	Gel
Amway	SA8 Fabric Softener Signature [13]	22.0	153	129	105	Gel

The only relatively concentrated product without high temperature issues is the Unilever Comfort Ultra. Analyses verified it is formulated with DHTDMAC, a material not considered to be biodegradable. All of the products based on esterquat developed significant changes in viscosity over time.

5.4 Conclusion

The rinse added fabric softener category has steadily moved to increasingly greener chemistry; however, there is much opportunity for improvement. This is a relatively high-volume category with increasing sales. Formulators will continue to seek improved stability with green softener actives (preferably based on plant feed stocks instead of animal) that can be further concentrated to reduce dosage and package needs.

References

[1] Euromonitor international – key strategies for growth in laundry care. August, 2010.
[2] Linfield WM, Sherrill JC, Davis GA, Raschke RM. Fabric treatment with cationic softeners. J Am Oil Chem Soc. 1958;35:590.
[3] Berenbold H. Utility and bio-behavior of rinse-added fabric softener. Tenside Surfactants Deterg. 1990;27:34.
[4] Levinson MI. Rinse-added fabric softener technology at the close of the twentieth century. J Surfactants Deterg. 1999 April;2(2):227–231
[5] Official Journal of the European Union; Regulation (EC) No 648/2004 of the European Parliament and of the Council of 31. March, 2004 on detergents.

[6] Mishra S, Tyagi VK. Ester quats: the novel class of cationic fabric softeners. J Oleo Sci. 2007;56:274
[7] Comfort Happy is a trademarked product of Unilever (Malaysia) Holdings SDN BHD.
[8] Hygiene Sweet is a trademarked product of the I.P. Manufacturing Company, Limited, Thailand
[9] Petals Superconcentrated Brightening is a trademarked product of the United Detergent Industries SDN BHD, Malaysia.
[10] Downy Nature Fresh is a trademarked product of the Procter & Gamble Company, Malaysia.
[11] Comfort Ultra is a trademarked product of Unilever (Thailand) Limited.
[12] Soflan Ultra is a trademarked product of the Colgate Palmolive Company, Malaysia.
[13] SA8 Fabric Softener Signature is a trademarked product of Amway

Raymond Neff and Agnieszka Gajewski

6 Renewable building blocks for polyurethanes

Abstract: For the past 60+ years, polyurethane chemistry has been used to make a wide variety of everyday consumer products such as mattresses, automotive interior parts and foam insulation. Today, the vast majority of polyurethane products are made entirely from petroleum. Although polyols made from natural oils have made inroads during the past decade, cost and performance, as well as the presence of a large-scale established infrastructure around petroleum-based materials, remain as barriers to significant market penetration. Promising new developments such as sugar-derived raw materials have the potential to shift the balance of cost and performance, but they are at an early stage. While not a comprehensive review of the large volume of academic literature on renewable polyurethane chemistry, this article discusses several routes to renewable, commercially viable building blocks for polyurethanes, considering both established and emerging technology.

Keywords: Polyurethanes, polyols, isocyanates, sustainability, renewable raw materials, natural oil polyols, sugar, isosorbide, carbon dioxide, eco-efficiency analysis

6.1 Polyurethane chemistry and applications

Ever since Otto Bayer filed the first polyurethane patent in 1937, polyurethanes have found use in a wide range of applications [1, 2]. Polyurethane foam is used for cushioning, insulation and automotive interior applications. Polyurethane elastomers are commonly used in footwear and sporting goods. The chemistry can also be used to formulate coatings, adhesives and sealants. Thermoplastic polyurethanes (TPU) can be processed with traditional extrusion equipment into tubing, wire coating, film or sheet. TPU can also be injection molded into parts such as shoe soles, automotive interior trim, caster wheels or mobile phone cases.

Polyurethanes are the product of diisocyanate and a polyol [1–3]:

$$\mathrm{OCN-R1-NCO\ +\ HO-R2-OH \rightarrow OCN-R1-NCOOR2-OH}$$

The polyol typically has a molecular weight from 200 to 6000 and may have a functionality (number of hydroxyl groups) between 2 and 8 depending on the application.

This article has previously been published in the journal Physical Sciences Reviews. Please cite as: Neff, R., Gajewski, A. Renewable Building Blocks for Polyurethanes. *Physical Sciences Reviews* [Online] **2018**, *3*. DOI: 10.1515/psr-2017-0193

https://doi.org/10.1515/9783110562781-006

Flexible foam used in bedding, furniture upholstery and automotive seating, requires a polyol with functionality around 3, and equivalent weight (per hydroxyl group) typically between 1000 and 2000. The backbone of such polyols is typically polyether or aliphatic polyester. The backbone can be a polyether (mixture of propylene oxide and ethylene oxide) or aliphatic polyester. If the foam is molded, a higher reactivity polyol, end-capped with ethylene oxide (producing primary hydroxyl groups) is typically used. Either MDI or TDI is typically used as the isocyanate. Water is added to the formulation to produce CO_2 gas, which expands the foam (with an amine as the intermediate):

$$2\ \mathrm{OCN-R1-NCO} + \mathrm{H_2O} \rightarrow \mathrm{CO_2} + \mathrm{OCN-R1-NHC(O)NH-R1-NCO}$$

Formulations also contain a nonionic silicone surfactant to stabilize the foam bubbles, and one or more catalysts (most commonly tertiary amines and/or organotin compounds).

Rigid polyurethane foam is commonly used as insulation for appliance or construction applications [1, 2]. Polyols for rigid foam are typically polyether or aromatic polyester. The polyether polyols are typically lower equivalent weight and/or higher in functionality than those used in flexible foam. Sorbitol or sucrose-initiated polyethers are typically the main polyol component (leading to theoretical functionalities of 6 or 8), although low equivalent weight diols and triols can also be added to optimize processing. Polymeric MDI is most commonly used as the isocyanate. Water can be used as a chemical blowing agent as in flexible foam, but physical blowing agents such as hydrocarbons and hydrofluorocarbons are more commonly used, in order to achieve low thermal conductivity. Other components such as silicone surfactants and tertiary amine catalysts are typically added.

Polyurethane elastomers are formulated primarily from polypropylene oxide diols ranging in molecular weight from 400 to 3000 [1, 2]. Polytetrahydrofuran, aliphatic polyesters (typically from adipic acid and butanediol or ethylene glycol), or polycaprolactone are also commonly used. Higher functionality polyols can be added in small amounts to achieve crosslinking. Low equivalent weight diols (e. g. ethylene glycol, propylene glycol or 1,4-butanediol) are added as "chain extenders", which react with the diisocyanate to form "hard segments." During polymerization, the hard segments separate from the polyol "soft phase", which becomes a physical crosslink in the material. Mechanical properties including hardness and tensile strength are adjusted with the polyol, crosslinker and chain extender. Polyurethane elastomers are most commonly made with 4,4'-MDI. However, aliphatic isocyanates (e. g. hexanediisocyanate, H_{12}-MDI or isophorone diisocyanate) can also be used, particularly if discoloration due to exposure to ultraviolet light is a concern.

TPU is typically formulated from diols, chain extenders and 4,4'-MDI [1, 2]. To produce a linear polymer, all components must have functionality close to 2.0. The diols are typically aliphatic polyesters and polytetrahydrofuran, with equivalent weight between 200 and 1500. Ethylene glycol, 1,3-propanediol, 1,4-butanediol and 1,6-hexanediol are typically used as chain extenders. Aliphatic isocyanates (e. g. hexanediisocyanate, H_{12}-MDI or isophorone diisocyanate) can also be used. TPU is manufactured in a one-shot continuous process, using either a conveyor belt or reaction extruder [2].

6.2 Traditional intermediates for polyurethanes

Over 90 % of polyols used for polyurethanes are polyethers [1]. Flexible and rigid polyurethane foam use polyethers incorporating propylene oxide (PO) and/or ethylene oxide (EO). These are produced using an alkoxylation process [2, 4]:

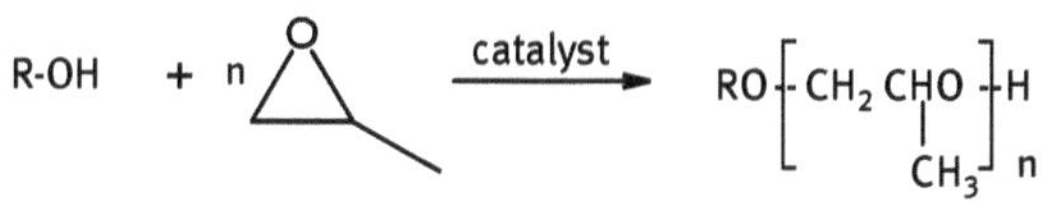

The polyol will have the same functionality as the initiator (i. e. 2 hydroxyl groups for propylene glycol, or 3 for glycerine). A mixed feed of PO and EO produces a statistical copolymer, while a sequential feed results in a block structure. Primary hydroxyl groups are formed if the end group is EO, and secondary hydroxyl groups result if the end group is PO. Primary hydroxyl groups provide faster reactivity and is desirable for molded articles (e. g. seat cushions and automotive trim parts), where cycle time is critical to manufacturing cost.

Polytetrahydrofuran (also known as Poly THF or PTMEG) is commonly used in cast elastomers and TPU. It is produced from the cationic polymerization of THF [5]:

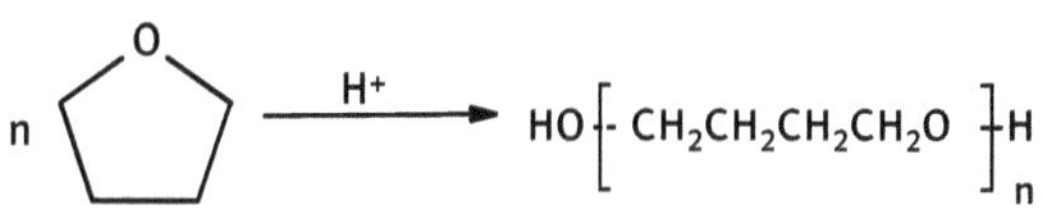

Aliphatic polyesters are produced from diacids such as adipic acid, and short diols such as ethylene glycol, 1,4-butanediol or 1,6-hexanediol [4].

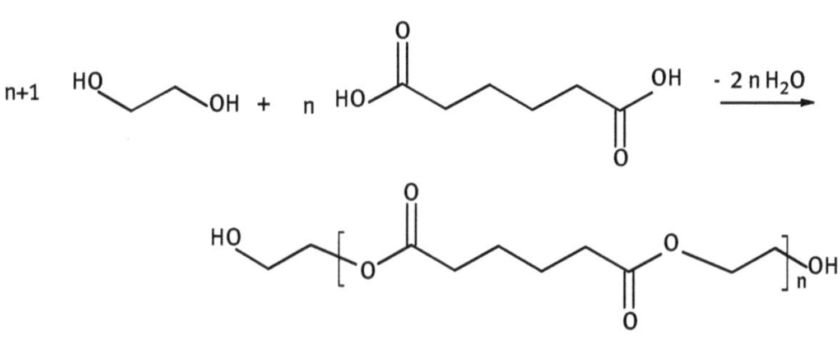

Aromatic polyesters (for rigid foam) can be produced from copolymerization of a glycol with aromatic acid derivatives, such as phthalic anhydride or dimethyl terephthalate:

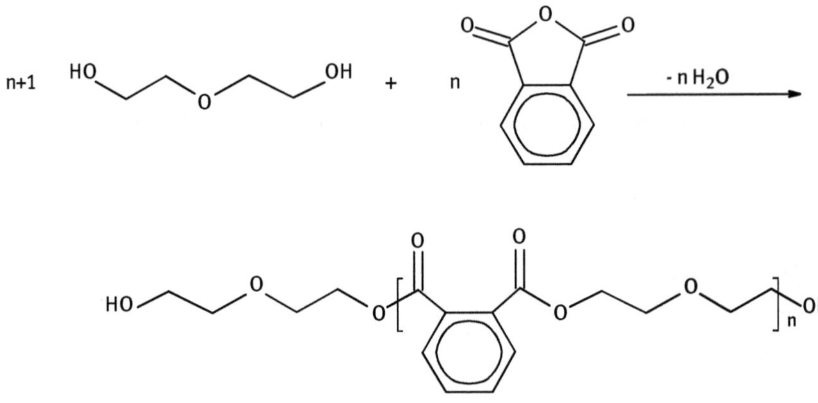

Polyester diols for can also be produced from caprolactone, a cyclic ester.

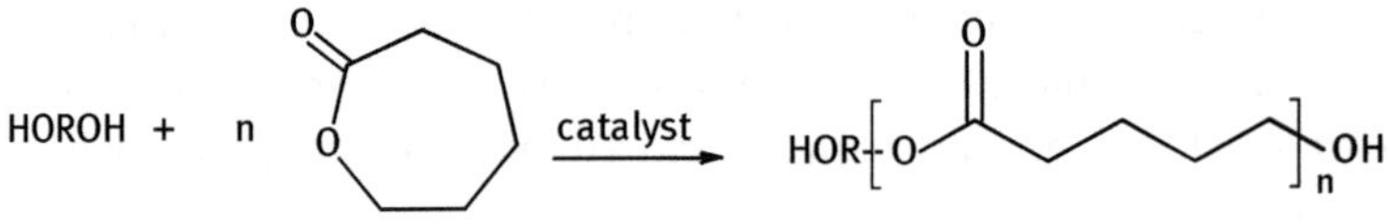

A polyester polyol with a narrower molecular weight distribution is produced. Polycaprolactone polyols are used in cast elastomer or TPU applications.

Hydrophobic polyols can be produced from butadiene. Polyurethanes made from these polyols exhibit improved moisture resistance, increased electrical resistivity and lower solid density than conventional polyurethanes. The resulting polyol is a mixture of structures formed by either 1,4-cis, 1,4-trans, or 1,2-addition of polybutadiene [4]:

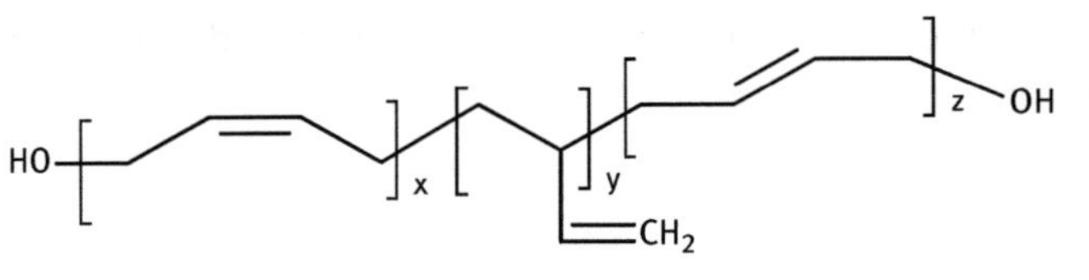

Free radical polymerization of butadiene, using a hydroxyl-containing initiator such as hydrogen peroxide, results in mostly 1,4-trans addition of butadiene, and 2.2–2.6 primary (mostly terminal) hydroxyl groups per chain. Commercial examples include Poly bd® from Total Cray Valley and Polyvest® HT from Evonik. Anionic polymerization of butadiene, followed by reaction with ethylene or propylene oxide, results in mostly 1,2-addition of butadiene, a more linear structure, narrow molecular weight distribution, and a functionality that is very close to the theoretical value of

2 hydroxyl groups per chain [4]. A commercial example is Krasol® from Total Cray Valley.

Since only one double bond reacts during the polymerization, polybutadiene polyols will contain double bonds in the backbone. The remaining double bonds are then available for crosslinking, either thermally, or by using sulfur or peroxide. Alternatively, the polyol can be hydrogenated in a subsequent step to saturate the double bonds.

Common chain extenders such as ethylene glycol, propylene glycol, diethylene glycol and dipropylene glycol are produced from ethylene or propylene oxide. 1,3-propanediol, 1,4-butanediol and 1,6-hexanediol are produced from other petroleum feedstocks.

MDI and TDI are the most commonly used isocyanates. MDI is produced from phosgenation of MDA, which is produced from aniline and formaldehyde [1, 2]. The process produces a distribution of oligomers and isomers:

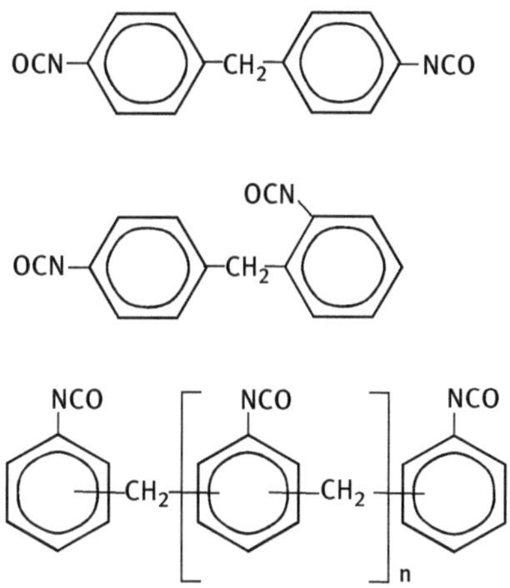

4,4’-MDI, 2,4’-MDI, and oligomers of functionality 3 and higher are the primary products, which are refined by distillation. Polymeric MDI is a mixture of monomeric MDI, and the higher oligomers. Different grades with various oligomer distributions are available.

TDI is produced from phosgenation of TDA, which is produced from toluene with dinitrotoluene as an intermediate. The process produces a mixture of 2 isomers, 2,4-TDI (80 %) and 2,6-TDI (20 %) [1, 2]:

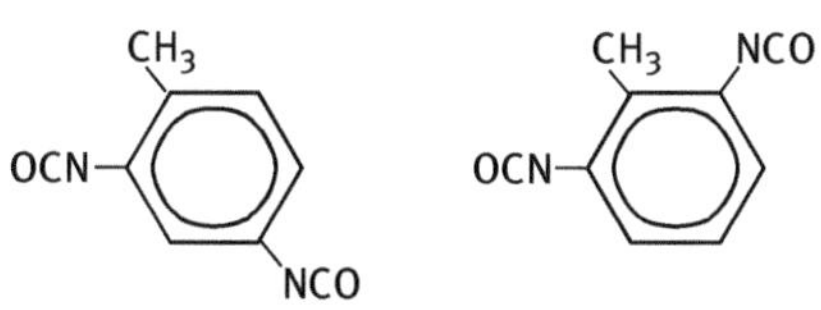

(80%) (20%)

6.3 Polyols and isocyanates from natural oils

Natural oils, such as vegetable, castor and palm oils, are good sources of building blocks for renewable polymers. These oils are typically triglycerides (esters of fatty acids and glycerine). Reactive sites in the oils are mainly carbon–carbon double bonds and ester groups, which are then converted to the desired functional groups (e. g. hydroxyl groups to produce polyols). This can be achieved through double bond reactions such as epoxidation, hydroformylation, metathesis, or through ester bond reactions (e. g. transesterification or transamidation) [6]. Since most oils contain multiple reactive sites, the resulting polymer will be crosslinked, leading to a thermoset. Therefore, the most common applications of vegetable oil based polyurethanes are rigid and flexible polyurethane foams.

When functional groups are introduced at the position of double bonds, the distribution of functional groups and structures is preserved. Therefore, hydroxyl groups introduced in the vegetable polyols are not terminal as in petroleum-based polyols. This leads to a network with dangling chains, which act as internal plasticizers and lower the Tg of the polymer [6]. Although the resulting network structure has a shorter distance between crosslinks, the presence of the dangling chains is likely beneficial in viscoelastic foam.

Soybean oil is most commonly used in North and South America, palm oil is used in Asia and rapeseed oil is used in Europe. Soybean oil is a triglyceride which contains 4.6 double bonds per molecule on average. The structure of soybean oil may vary from molecule to molecule, since it combines five different fatty acids: linoleic, linolenic, oleic, palmitic and stearic acids.

Linoleic Acid	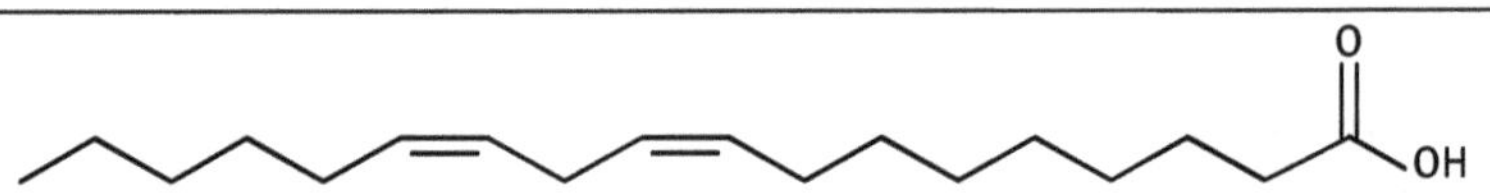
Linolenic Acid	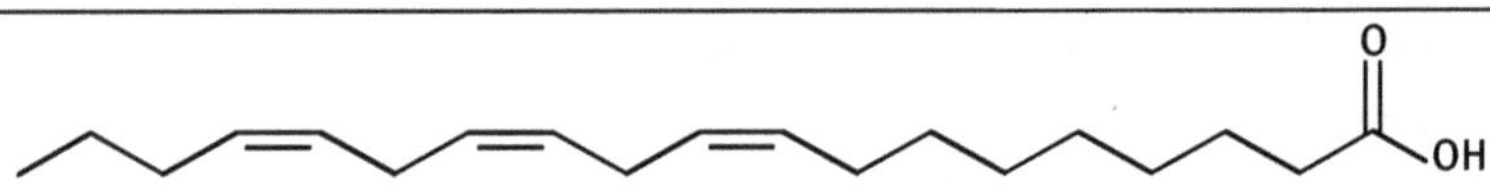
Oleic Acid	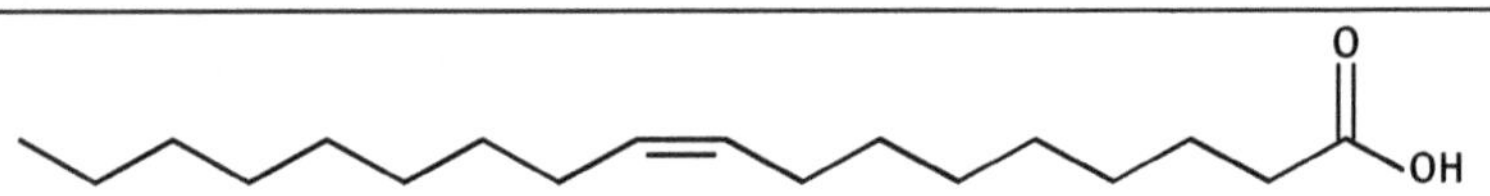
Palmitic Acid	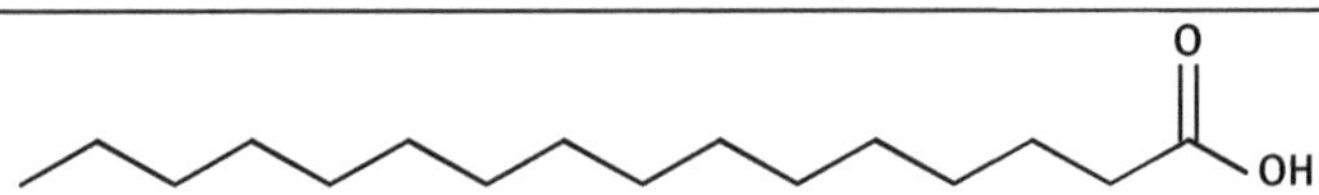
Stearic Acid	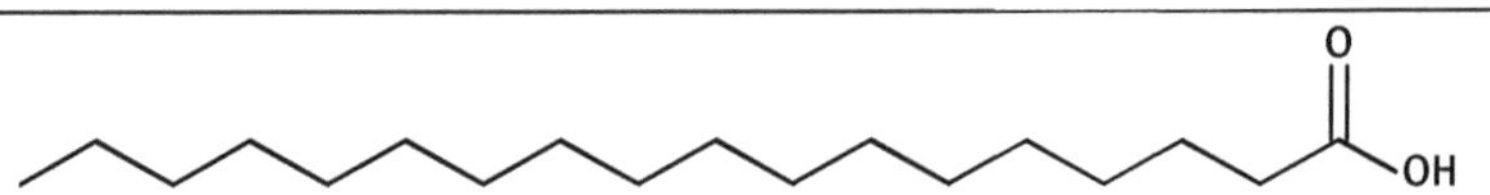

Although direct oxidation of double bonds introduces hydroxyl groups to the vegetable oil, controlling the process is challenging due to the many byproducts formed (i. e. peroxides, aldehydes or ketones). These byproducts often give rise to odor, lower functionality or higher acid value requiring additional post processing and increasing cost. Epoxidation of the double bonds is one alternative route. This can be achieved by catalytic hydrogenation, reaction with hydrochloric or hydrobromic acid and by acid catalyzed ring opening [6]. However, a range of polyol structures is produced that may impart different properties to the final polyurethane.

Commercial soybean oil polyols include Renuva™ polyols sold by Dow, and BiOH™ polyols sold by Cargill. A commercial scale process to produce soybean oil polyols has been reported by Dow [7]. In the first step, soybeans are extracted with methanol to produce fatty acid methyl esters. Hydroformylation of double bonds in the soybean oil and subsequent reduction produces aldehydes and alcohols. A catalyst (most commonly rhodium or cobalt carbonyl) is required for hydroformylation reactions. Total recovery of the catalyst is required to keep the process cost-effective. The monomers are then polymerized with an initiator to prepare the final polyol through a transesterification reaction. The resulting polyol includes an extra carbon at every double bond, resulting in longer chains, and ultimately softer and more elastic materials [6].

Methanolysis:

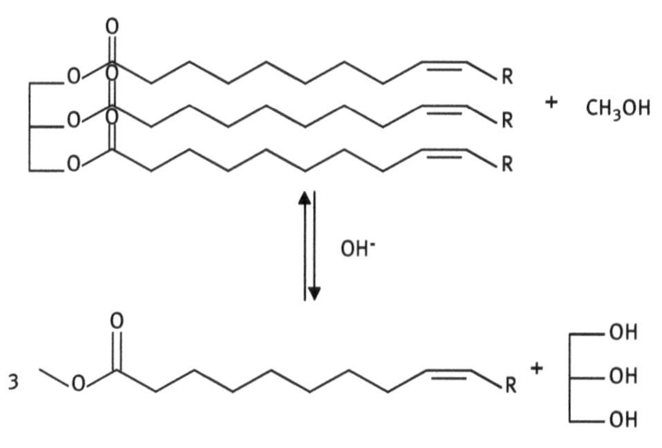

Hydroformylation:

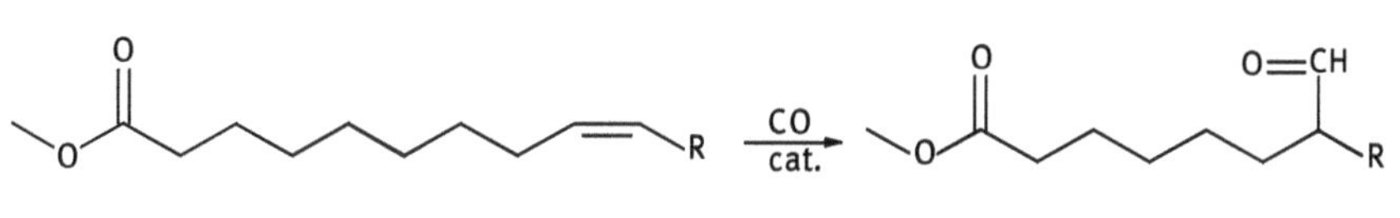

Hydrogenation:

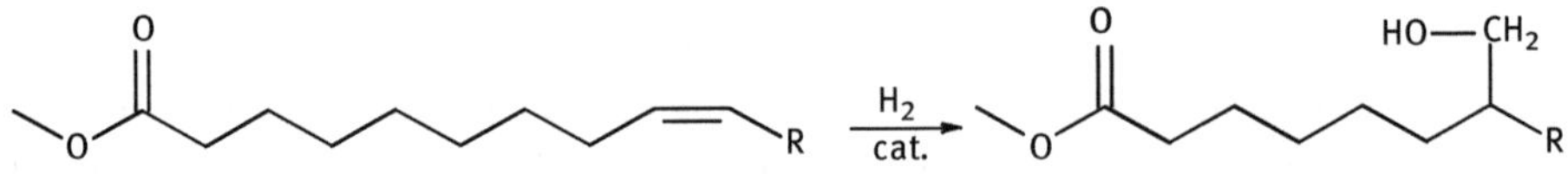

Polymerization:

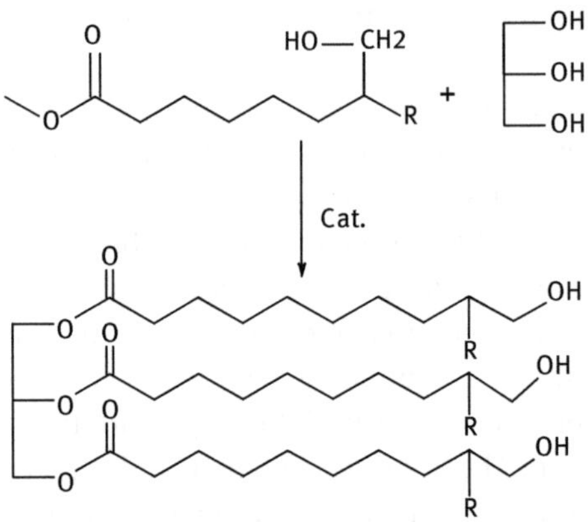

Castor oil is a triglyceride containing 90 % ricinoleic acid, which has one hydroxyl group on the 12th carbon, and a double bond between the 9th and 10th carbon.

Ricinoleic Acid:

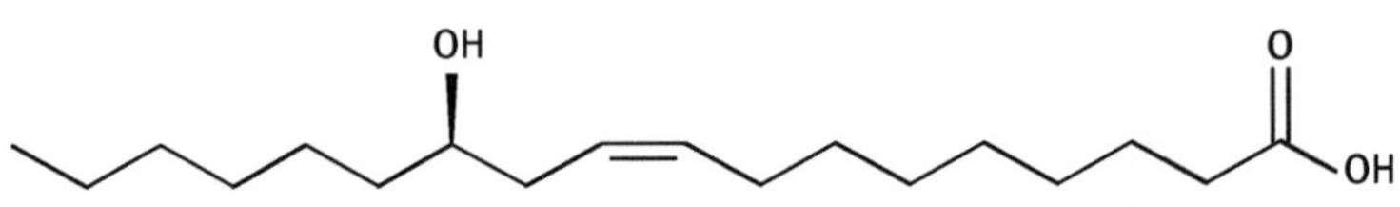

Unmodified castor oil can be used as a urethane polyol [6, 8]. Although it has a hydroxyl functionality of 2.7, its hydrophobicity and molecular weight of approximately 1000 restrict its use in flexible foam. A limited amount can be added to flexible or rigid foam formulations, but will have a negative effect on properties and processing. Smiecinski et al. [8] describe a propoxylated castor oil polyol (marketed by BASF as Pluracol® Balance™50), which addresses these issues. Castor oil accounts for 31 % of the polyol, and foams made from the polyol contain up to 20 % renewable content. The propylene oxide in the polyol backbone provides sufficient molecular weight and compatibility with the other formulation components, in contrast to castor oil and most other natural oil polyols. Therefore, the polyol can be used as the sole polyol in flexible slabstock foam with minimal changes in formulation, processing or physical properties, compared with foam made using a standard petroleum-based polyol.

Bradlee et al. conducted a life cycle analysis of castor oil/propylene oxide polyols for flexible slabstock foam using an internal eco-efficiency analysis [9]. Comparisons were made to the standard petroleum-based polyol. Overall costs, primary energy consumption, ozone depletion potential, acidification potential, photochemical ozone creation (smog), water emissions, solid waste generation, raw material consumption, toxicity potential, risk potential and land use were considered. In general, the environmental impact of the castor oil/PO polyol depended on its molecular weight. The lower molecular product was more viscous and required heat in order to pump the product, and this affected the analysis. Overall, the high MW castor oil/PO product was the most eco-efficient.

Dimerized fatty acids, also known as "dimer diacids", are produced from natural oils such as soybean oil or tall oil. They are obtained by processing a mixture of fatty acids at elevated temperature and pressure using a catalyst, followed by distillation and hydrogenation [10, 11].

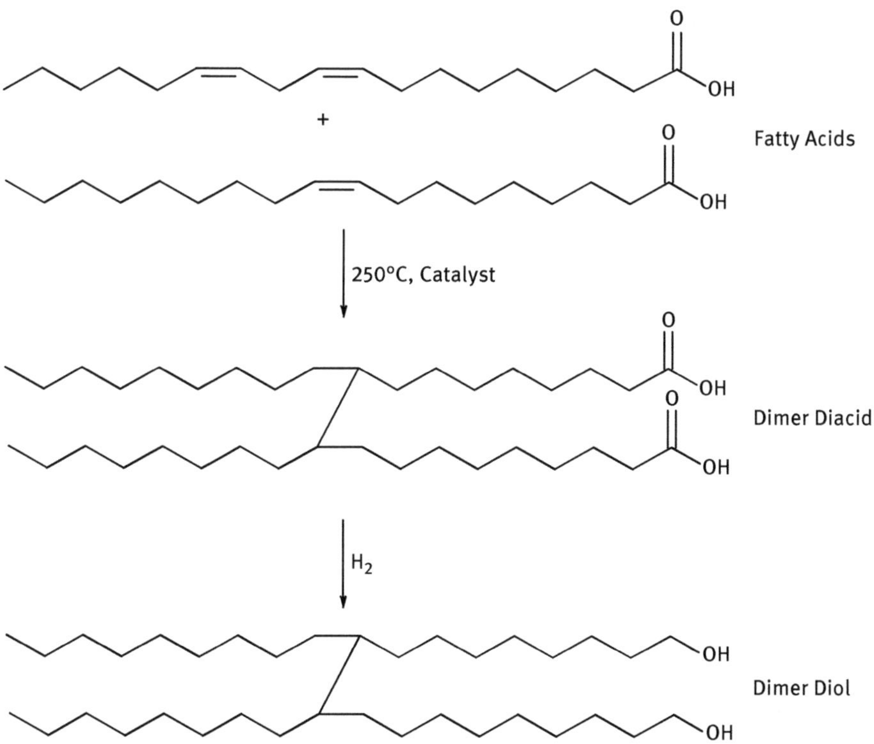

The resulting 36-carbon, di-functional acid can be converted into the corresponding diol, or can be chain extended with a short chain diol (e. g. ethylene glycol), and optionally another diacid (e. g. adipic acid) to produce a polyester diol [10].

Dimer-based polyols are sold under brand names such as Sovermol® (BASF), Priplast™ and Pripol™ (Croda). The dimer-based polyols were shown to produce polyurethane elastomers, which have comparable hydrolysis resistance to PTMEG, and are significantly more hydrolysis resistant than for typical polyester diols (e. g. hexanediol-adipate) [10]. Moreover, the polyurethanes from the dimer-based diols exhibit less moisture uptake than those from PTMEG, or from any other typical polyether or polyester polyol. The dimer-diol polyesters are also more resistant to heat, oxidation and UV attack since they do not contain ether linkages. In addition to elastomers, dimer-based polyols can be used for coatings, adhesives and flexible foams.

A diisocyanate based on a dimer diacid, DDI 1410, is sold by BASF [6, 12]. The product is used mainly in coatings to achieve increased flexibility and moisture resistance. It is also less toxic than typical aliphatic diisocyanates.

6.4 Sugar as a source for polyols and chain extenders

DuPont Tate&Lyle produces Susterra® propanediol from corn sugar using a proprietary process [13]. The material can be used as a chain extender in TPU or polyurethane elastomers, replacing 1,4-butanediol, or petroleum-based 1,3-propanediol. DeSalvo et al. [13] compared the performance of Susterra® Propanediol to 1,4-butanediol in MDI and TDI polyurethane elastomers made by the prepolymer method. Physical property performance was comparable, with improvements in abrasion and increased pot-life in some cases.

Susterra® propanediol can also be reacted with adipic acid, or another suitable diacid to produce an aliphatic polyester polyol. Use of such a polyol in combination with the renewably sourced diol as a chain extender increases the renewable content even further. DeSalvo et al. [13] evaluated adipates made with Susterra® propanediol of molecular weight between 500–2000 Da in MDI elastomers using the prepolymer method. Use of Susterra® propanediol as the chain extender has resulted in elastomers with up to 55 % renewable content. Overall properties of the studied elastomers were comparable with industry standards, with some improvements reported in resilience and abrasion.

By polymerizing the bio-sourced propanediol in the presence of a catalyst, a 100 %-biosourced polyether polyol is produced:

$$\text{n HO} - \text{CH}_3{-}\text{CH}_3{-}\text{CH}_3{-}\text{OH} \rightarrow \text{HO} - (\text{CH}_3{-}\text{CH}_3{-}\text{CH}_3\text{O})_n{-}\text{H} + \text{nH}_2\text{O}$$

The polyol, also referred as polytrimethylene ether polyol, PTMEPOL or “PO3G”, has been produced in molecular weights between 500 and 3000, and has been marketed under DuPont’s brand name Cerenol™ [14]. TPU or polyurethane elastomers with similar properties to those based on polytetrahydrofuran can be produced using this polyol.

Isosorbide (1,4:3,6 diandydrosorbitol or dianhydro-D-glucitol) is a diol which is formed from the double dehydration of sorbitol (sorbitol can be obtained from hydrogenation of glucose) [15]:

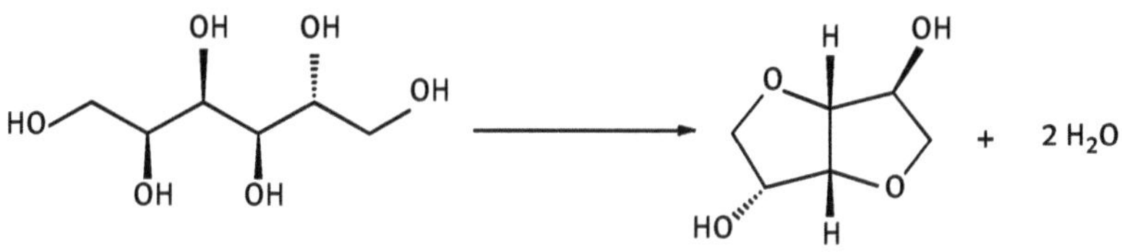

Due to the rigidity of the molecule, it is a potential alternative to Bisphenol A. Unfortunately, the two hydroxyl groups do not have equivalent reactivity, since the molecule forms an assymetric planar structure, resulting in one of the hydroxyl groups hydrogen-bonding with the neighboring ether oxygen [15]. This creates difficulty in using the molecule as a chain extender for elastomers, or into polyester polyols, since the different reactivity will inhibit molecular weight build. However, use of isosorbide in polyurethanes is still possible though appropriate use of catalysts and elevated reaction temperature. Additionally, isosorbide is not soluble in typical flexible polyether polyols. Capping isosorbide with ethylene oxide or propylene oxide could result in a material suitable for use in elastomers or TPU.

Ionescu et al. [15] demonstrated the use of isosorbide in rigid polyurethane foam. Isosorbide was successfully added directly to the foam formulation by dissolving it in a sucrose-initiated polyether polyol, and the foam had properties comparable with typical rigid foam.

Schneiderman et al. [16] reported another route to polyurethanes using sugar as a starting material, producing foams and elastomers with similar properties to petroleum-based polyurethanes. A polyol was synthesized in the bulk from β-methyl-δ-valerolactone (MVL).

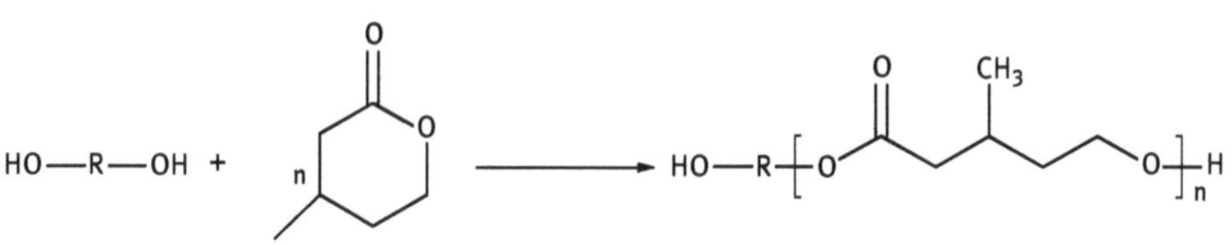

MVL can be produced from sugar, at potentially low cost. The backbone of the polyol (pMVL) resembles polycaprolactone, which is currently used as a building block for polyurethane elastomers. Furthermore, the polymer degrades back to monomer upon heating between 200 and 250 °C. The MVL monomer can be recovered through distillation, resulting in the added benefit of closed loop recyclability for this material. Chemical recycling of typical polyurethanes (through hydrolysis or glycolysis) is not straightforward and yields a mixture of recycled polyols with limited utility.

Yoo et al. [17] reported a route to unique sugar based polyurethanes using poly(trans-β-farnesene) diols, developed by Amyris and Total SA. The monomer, β-farnesene (BioFene™), is produced by yeast fermentation of sugar syrups.

Hydrophobic diols (with characteristics comparable with polybutadiene diols) are produced by anionic polymerization and can be used for cast elastomers or thermoplastic polyurethanes. The resulting polymer diols have a narrow molecular weight distribution and a dense "bottlebrush" structure:

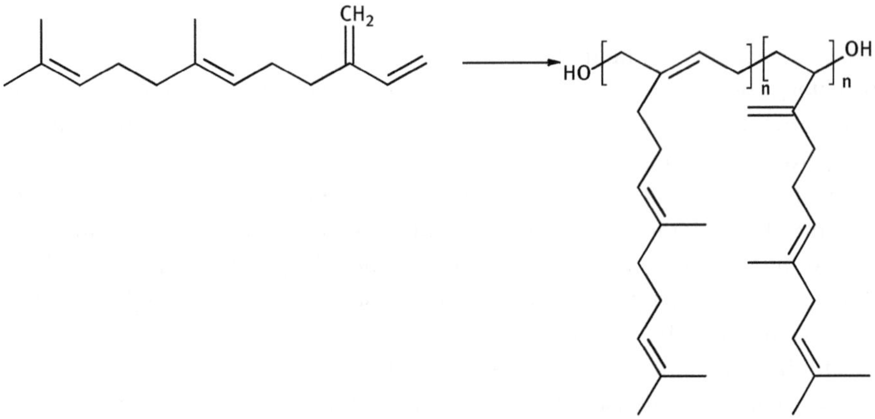

The dense side-chain structure inhibits chain entanglement, even at high molecular weight. The viscosity of the polymer is lower than a comparable polybutadiene. Hydrogenation of the structure leads to tighter chain packing, reduced free volume and higher Tg. Properties of polyurethane elastomers based on poly(trans-β-farnesene) diols were not reported.

6.5 Carbon dioxide as a building block (polyether carbonates)

Carbon dioxide can be incorporated into polyols through copolymerization with propylene oxide using a double-metal cyanide catalyst to form an alternating polyethercarbonate with propylene carbonate as a by-product [18, 19]:

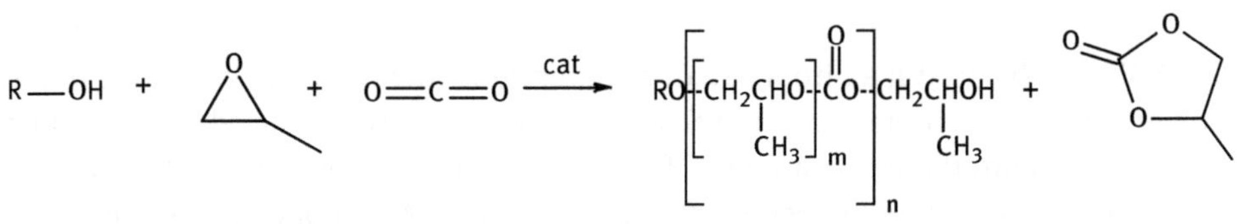

While not bio-sourced, CO_2 is a renewable material, available from the atmosphere, or from a fossil fuel power plant with carbon capture. A lifecycle analysis [18] indicated that incorporating 20 wt% CO_2 reduces greenhouse gas emissions by 11–19 % and reduces fossil resources by 13–16 % compared to production of conventional polyether polyols. However, the process does emit a net positive amount of CO_2, and thus is not a sink for greenhouse gas.

Since the polyol has relatively low viscosity and produces polyurethanes with low glass transition temperature, it is suitable for use in flexible polyurethane foam. Physical properties of flexible polyurethane foams produced using a polyethercarbonate polyol are comparable to those produced using conventional polyether polyols [19].

6.6 Future outlook

Starting with products derived from vegetable oils, renewable materials have made inroads as building blocks for polyurethanes. However, most polyurethane will continue to be made from petroleum for the near future. Issues to overcome are primarily the cost of the renewable materials compared with their performance, as manufacturers and consumers are generally not willing to pay simply for renewable content. Many of the most recent innovations are from sugar-derived materials. New innovations will continue to drive the use of renewable materials in polyurethanes, especially where unique properties are obtained, and when economies of scale can be reached in order to lower costs.

References

[1] Woods G, editor. The ICI polyurethanes book. 2nd ed. New York: John Wiley & Sons, 1990.

[2] Oertel G, editor. Polyurethane Handbook. Munich: Hanser Verlag, 1985.

[3] Saunders H, Frisch KC. Polyurethanes: chemistry and technology, part I, chemistry. New York: Interscience Publishers, 1962.

[4] Ionescu M, editor. Chemistry and technology of polyols for polyurethanes. UK: Rapra, 2016.

[5] Pruckmayr G, Dreyfuss P, Dreyfuss MP. Polyethers, Tetrahydrofuran and Oxetane Polymers" in Kirk-Othmer Encyclopedia of Chemical Technolog. New York: John Wiley & Sons, Inc., 1996.

[6] Petrovic ZS. Polyurethanes from Vegatable Oils. Poly Rev. 2008;48:109–55.

[7] Babb D, Phillips J, Keillor C, 2006. Soy-based polyol for flexible slabstock foam. In: Proceedings of the API Polurethanes 2006 Technical Conference, September 25–27, 2006, Salt Lake City, UT.

[8] Smiecinski T, Manea V, Langer DT, 2008. Renewable polyols for polyurethane flexible foam. In: Proceedings of the Center for the Polyurethanes Industry Technical Conference, San Antonio, TX Sept. 29–Oct. 1, 2008.

[9] Bradlee CA, Manea VP, Kloock LM. Eco-efficiency analysis demonstrates the environmental benefits of flexible furniture foams made with BALANCE™ 50 polyol. In: Proceedings of the Center for the Polyurethanes Industry Technical Conference, San Antonio, TX, Sept. 29-Oct. 1, 2008.

[10] Appelman E, Danek B, Van Der Wouden M, 2000. Dimerized fatty acid technology in polyurethanes. In: Proceedings of Polyurethanes Conference 2000, Boston, MA, Oct. 8–11, 2000.

[11] Hill K. Fats and Oils as Oleochemical Raw Materials. Pure Appl Chem. 2000;72:1255–64.

[12] Liu Z, Kraus G, editors. Green Materials from Plant Oils. Cambridge, UK: Royal Society of Chemistry, 2014.

[13] DeSalvo JW, Miller ES, Miller R, Van Gorp JJ, Franyutti S, 2009. Susterra® propanediol – from the leading edge of biotechnology, a renewable and sustainable building block for the urethane

industry that delivers performance and versatility. In: Proceedings of Polyurethanes 2009 Technical Conference, Washington, D.C., Oct. 5–7, 2009.

[14] Sunkara HB, Miller RW. DuPont™ Cerenol™. A new family of high performance polyether glycols. Fifth Annual World Congress on Industrial Biotechnology & Bioprocessing, April 27-30, 2008, Chicago, IL.

[15] Ionescu M, Petrovic ZS, Sandhu MD, Javni I, Bilic N, 2011. New isosorbide derivatives for bio-based polyurethanes. In: Proceedings of Polyurethanes 2011 Technical Conference, Nashville, TN, Sept. 26–28, 2011.

[16] Schneiderman DK, Vanderlaan ME, Mannion AM, Panthani TR, Batiste DC, Wang JZ, et al. Chemically Recyclable Biobased Polyurethanes. ACS Macro Lett. 2016;5:515–18.

[17] Yoo T, Henning SK. The synthesis and characterization of farnesene-based polymers. In: Fall 190th Meeting of Rubber Division, ACS, Cleveland, OH, Oct. 11–13, 2016.

[18] Von Der Assen N, Bardow A. Life cycle assessment of polyols for polyurethane production using CO_2 as feedstock: insights from an industrial case study. Green Chem. 2014;16:3272.

[19] Langanke J, Wolf A, Hofmann J, Böhm K, Subhani MA, Müller TE. Carbon dioxide (CO2) as sustainable feedstock for polyurethane production. Green Chem. 2014;16:1865 .

Pei-Yu Kuo, Ning Yan, Nicole Tratnik and Jing Luo

7 Applications of bark for bio-based adhesives and foams

Abstract: With the increased concern for climate change and depletion of fossil fuel resources, there is a growing trend to research and develop technologies that can use renewable biomass as the raw material for synthesizing chemical products. Bark, a largely available forestry biomass residue with attractive chemical compositions, is considered as a promising feedstock. This article summarizes our recent research and development work in deriving bark-derived adhesives and foams and various bark conversion technologies explored. Advantages and disadvantages associated with the conversion technologies and bark-based chemical products are discussed. Some future studies that can further promote commercial applications of these novel bio-based products are presented. These novel bark-derived products have potential to generate higher value return using the low-valued forestry residue materials while increasing the renewable content in the final chemical products for a higher sustainability.

Keywords: Alkaline Extraction, Solvolytic Liquefaction, Alkoxylation, Phenol-Formaldehyde Resins, Melamine-Formaldehyde Resins, Epoxy Resins, Polyurethane Foams

> *"Fire made us human, fossil fuels made us modern, but now we need a new fire that makes us safe, secure, healthy and durable."*
>
> – *Amory Lovins*

7.1 Barkvalorization toward a more sustainable future

Due to the concern of climate change and the dwindling of fossil fuels, developing new materials with higher sustainability has become a necessity. Sustainability is emerging as a fundamental requirement in the design of novel products. To achieve environmental sustainability, renewable feedstock plays a crucial role in reducing carbon footprint and achieving greenhouse gas mitigation at the lowest possible cost. Bark, as a promising biological resource that can be used as renewable feedstock for chemical synthesis, has received extensive attention from the government,

This article has previously been published in the journal Physical Sciences Reviews. Please cite as: Kuo, P.-Y., Yan, N., Tratnik, N., Luo, J. Applications of Bark for Bio-Based Adhesives and Foams. *Physical Sciences Reviews* [Online] **2018**, *3*. DOI: 10.1515/psr-2017-0194

https://doi.org/10.1515/9783110562781-007

academia, and industry. Thus, this review summarizes our recent research and development work in deriving chemicals from tree bark, i. e. a sustainably sourced biomass residue material. These novel bark-derived chemical products have potential to generate higher value return using waste residue from the timber industry and to reduce the usage of toxic synthetic chemicals, such as phenol and bisphenol A (BPA), in their synthesis.

7.1.1 The rationale for bark

Bark, a non-food based biomass source, can be used to substitute for petroleum-based chemicals due to its abundance, accessibility, and molecular features. In addition to the environmental and technical advantages of using bark for chemical production, there are positive social and economic outcomes as well. The effects of converting bark into chemicals are summarized in Figure 7.1. The worldwide bark yield is roughly 60 million dry tons [1], but most of the tree barks are under-utilized and are either left in the field after harvest or burned as a low-efficiency fuel. Furthermore, there is an extra supply of tree barks in Canada due to the 20.3 million hectares of forest that were damaged by insects in 2014 [2]. In an attempt to salvage these dead trees, the forest sectors in Canada and the USA have invested billions of dollars to develop new bio-derived products. One of the most promising applications is to convert bark into high-value chemicals.

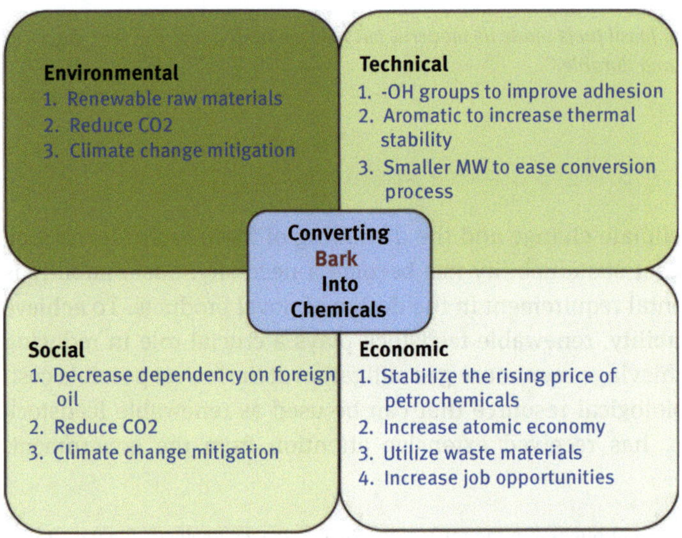

Figure 7.1: The social, economic, environmental, and technical benefits of converting bark biomass into high-value chemicals.

Bark-derived chemicals have been synthesized and used since ancient times. For instance, Egyptians applied bark-derived resins on coffins for mummies, and Asians painted latex found in the bark of rubber trees on lacquerware to form protective layers [3]. Before modern chemistry, many applications of bark had been discovered by chance or experimentation. Around 1930, researchers started to have a more comprehensive understanding of bark and produced bark-derived products in a systematic manner on an industrial scale. For example, ITT Rayonier Inc. developed highly reactive bark extractives to be combined with formaldehyde for adhesive applications in 1977 [4], and a large number of bark extraction technologies and bark-based products were patented from 1955 to 1980 [5–10]. The studies of bark on polymer synthesis did not last for long, as synthetic chemicals have dominated the polymer market since the mid-twentieth century. During that time, the usage of bark was fairly limited to low grade fuel for energy generation. However, the high oxygen content of bark reduces its caloric value, which actually makes bark a low grade fuel for energy generation. In contrast, to maximize the efficiency of its utilization in chemical generation, the final products of bark derived chemicals should keep as much oxygen as possible. The oxygen remains in bark or bark extractives as hydroxyl groups that can improve the adhesion of polymers to metals or can be used for producing polyols in polyurethane foams. In fact, the two biggest potential applications of bark constituents are currently adhesives and foams.

The large number of aromatic groups and the lower average molecular weight of bark compounds compared to those in wood can benefit the application of bark in adhesives and foams. Higher amounts of aromatic groups can enhance the thermal stability of its derived products. Additionally, the natural role of bark is acting as a protective layer of wood against environmental degradations, which can degrade lignin into smaller fragments in the bark layer. These smaller molecular weight compounds can potentially facilitate the bark conversion process and subsequent manufacturing steps. It is well-known that molecular features have significant impacts on the properties of bark-derived chemicals, so to be able to successfully develop bark-containing products, it is crucial for researchers to have a comprehensive understanding of the chemical composition of bark. The following section provides a general understanding of the molecular composition of bark.

7.1.2 Chemical composition of bark

The chemical composition of bark can be divided into four categories: cellulose, hemicellulose, lignin, and extractives. Compared to wood, bark usually contains a higher percentage of extractives and lignin, as shown in Table 7.1 [11]. The chemical composition of bark is diverse and significantly dependent on the species, growing conditions, extraction methods, extraction temperature, and particle size; thus, the application of particular bark samples should also be adjusted according to their sources, structures, and compositions. In addition to these, many bark components may have been photo-degraded prior to timber harvesting, and it is the existence of these small, degraded molecules that facilitates their subsequent extraction and use as a feedstock to produce bio-based polymers.

Table 7.1: Percent composition between wood and bark in softwood and hardwood [11].

	Softwoods		Hardwoods	
	Wood	Bark	Wood	Bark
Lignin*	25–30	40–55	18–25	40–50
Polysaccharides*	66–72	30–48	74–80	32–45
Extractives	2–9	2–25	2–5	5–10
Ash*	0.2–0.6	Up to 20	0.2–0.6	Up to 20

* Based on the extractive-free material (Unit:%)

7.1.2.1 Polysaccharides

Cellulose and hemicellulose are the two major polysaccharide components. Cellulose is a linear polymer of D-glucose linked by β-1,4 bonds, and its degree of polymerization (DP) is usually above 5000. In contrast, hemicellulose is a branched polymer of various sugar units such as D-xylose, D-mannose, D-galactose, D-glucose, and L-arabinose, and its DP is usually below 300. Due to the structure and composition of hemicellulose, multi-sugar-based hemicellulose is more susceptible to hydrolysis than cellulose. During the extraction or liquefaction processes, both polysaccharides can decompose into oligosaccharides, monosaccharides, and pyrolysis products such as glyceraldehydes and 5-hydroxymethyl-2-furaldehyde. If the extraction conditions are even harsher than in those processes, more degraded compounds are generated, which will be discussed in the liquefaction section. The sugar percentages and sugar types of various bark can be found in Table 7.2 [11].

7.1.2.2 Lignin

Lignin is a complex, 3D amorphous cross-linked polymer that can be simplified into three basic units: syringyl, guaiacyl, and p-coumaryl alcohol. Syringyl is found in softwood, syringyl and guaiacyl are in hardwoods, and all three are present in grasses. These basic units are linked mainly by β-O-4 bonds (usually more than 50 % of all linkages). Other common bonds are 5–5, β-5, and α-O-4 linkages. Among all the linkages, β-O-4 bonds have a relatively low bond energy and is therefore one of the first bonds cleaved during the liquefaction process. The details of how β-O-4 bonds break into corresponding carboxylic acid and phenol monomers can be found in our previous research [12]. Additionally, it should be noted that lignin has a

Table 7.2: The sugar content and types of various bark [11].

Species	Reducing sugars from–		
	Unextracted bark	Extractive-free bark	Alkali-extracted bark[1]
	%	%	%
Fir, balsam	46.6	45.3	32.9
Larch, Western	46.6	46.0	38.0
Spruce, Engelmann	42.9	34.3	24.2
Spruce, black	47.9	44.8	32.3
Pine, jack	30.6	28.8	21.1
Pine, lodgepole	38.3	32.9	19.2
Pine, slash	29.7	29.8	26.4
Pine, sugar	22.1	19.8	16.1
Pine, Western white	42.6	34.0	26.0
Hemlock, eastern	34.9	33.3	29.1
Boxelder	40.6	37.8	30.0
Maple, sugar	35.4	34.3	31.1
Alder, red	38.6	38.0	30.3
Birch, yellow	32.5	31.8	26.0
Birch, paper	32.2	30.1	21.8
Pecan	33.5	30.7	23.3
Sweetgum	35.6	33.5	26.4
Blackgum	29.6	27.2	22.4
Sycamore, American	40.9	39.0	31.1
Cottonwood, swamp	41.0	39.2	34.1
Aspen, quaking	41.4	39.7	34.9
Oak, white	27.8	28.2	21.2
Oak, northern red	32.4	31.7	28.3
Willow, black	42.9	43.4	35.4
Elm, American	37.0	35.4	27.0

Species	Glucose	Galactose	Mannose	Arabinose	Xylose
	Percent	Percent	Percent	Percent	Percent
EXTRACTIVE-FREE BARKS					
Fir, balsam	64	5	12	9	7
Larch, western	69	4	11	6	9
Spruce, Engelmann	61	5	9	13	9
Spruce, black	64	6	7	11	9
Pine, jack	64	7	6	10	11
Pine, lodgepole	50	7	6	26	8
Pine, slash	63	7	7	7	15
Pine, sugar	69	6	8	7	9
Pine, Western white	75	3	6	2	15
Hemlock, eastern	67	3	13	8	7
Boxelder	65	3	2	7	20

(continued)

Table 7.2: (*continued*)

Species	**Glucose Percent**	**Galactose Percent**	**Mannose Percent**	**Arabinose Percent**	**Xylose Percent**
Maple, sugar	63	3	1	6	25
Alder, red	54	3	1	6	34
Birch, yellow	54	3	1	8	32
Birch, paper	53	2	1	6	36
Pecan	69	4	1	11	11
Sweetgum	60	3	3	11	20
Blackgum	60	4	1	8	24
Sycamore, American	59	4	1	4	30
Cottonwood, swamp	61	4	1	5	26
Aspen, quaking	60	2	1	5	30
Oak, white	60	4	2	7	24
Oak, northern red	53	3	1	6	35
Willow, black	69	3	1	6	18
Elm, American	70	4	2	9	11
ALKALI-INSOLUBLE BARK RESIDUES[1]					
Larch, western	74	4	9	4	9
Pine, slash	67	6	7	5	15
Pine, western white	74	2	3	3	17
Hemlock, eastern	76	2	10	5	7
Maple, sugar	69	2	1	4	24
Birch, paper	55	2	1	5	37
Sweetgum	71	2	1	5	21

high degree of hydroxyl functionality as primary, secondary, and phenolic alcohols. The phenolic part of the lignin molecule is a precursor for phenolic monomers or oligomers which can be used as substitutes for phenolic precursors in adhesives such as polyurethanes, epoxies, and phenol formaldehyde.

7.1.2.3 Extractives

The family of bark extractives can be roughly divided into four sub-categories: phenolic substances, terpenes/terpenoids, fats/waxes, and inorganic compounds (Figure 7.2). These compounds can be extracted from bark by using solvents as shown in Table 7.3 [11]. Typical solvents include aqueous sodium hydroxide solutions, simple alcohols, alkanes, benzene, diethyl ether, and acetone. Other factors that affect the compounds that are extracted are leaching conditions (e. g., temperature) and the botanical origin (being highly species and tissue dependent). The variation in extractive contents depending on species and on extraction solvent can be seen in Table 7.4 [11].

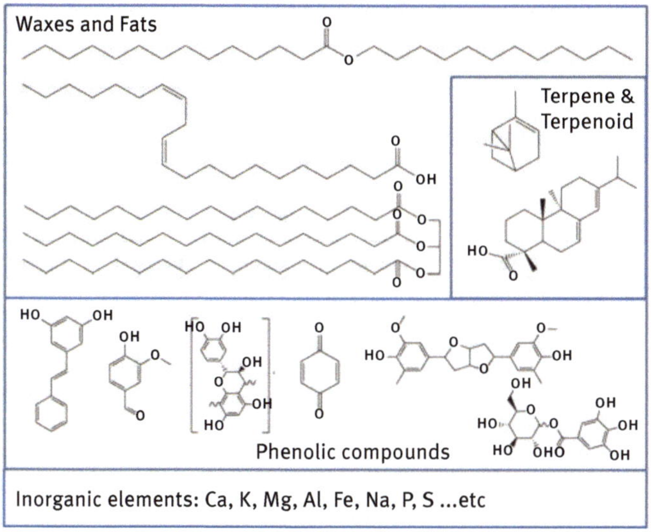

Figure 7.2: Chemical structures of extractive compounds from bark. Modified with permission from D'Souza J. The Utilization of Bark to Make Rigid Polyurethane Foams. Toronto: University of Toronto Thesis; 2016.

Table 7.3: Dependence of extractives composition on solvent [11].

Solvents	Type of extractive compounds
Petroleum ether, ether, benzene, chloroform	Terpenes, fats, waxes, fatty and wax acids, sterols, resins
Alcohol, acetone, aq. alcohol, aq. acetone	Simple polyphenols and their glycosides, tannins, mono and disaccharides
Hot/cold water	Disaccharides, starch, gums, pectins, tannins, mucilages
Aq. alkali	Phlobaphenes, phenolic acids, lignin, hemicellulose, suberin
Acid hydrolysis	Sugars, uronic acids

As mentioned above, the chemical composition of bark extractives can vary dramatically depending on the species, conversion technology, and solvent. Thus, it is important to choose the right origin and conversion procedure to produce the desired bark-based chemicals. For example, to extract polyphenols to use as a substitute for petroleum-derived phenolic compounds in the production of phenolic resins and rigid polyurethane foams, lodgepole pine would be an ideal raw material since it has the highest yield after alkaline extraction and second highest yield of formaldehyde-condensable polyphenols, as shown in Table 7.5 [17]. In order to explain this example further, this article focuses mainly on our own research work from 2010 to 2017 in

Table 7.4: Various extractions of barks (Percentages based on weight of oven-dried unextracted bark) [11].

Species	Material soluble in 1% NaOH[1]	Material dissolved by successive extractions with–			
		Benzene	95% alcohol	Hot water	1% NaOH
	Percent	Percent	Percent	Percent	Percent
Fir, balsam	49.4	13.2	3.3	2.7	30.6
Larch, western[2]	43.6	1.3	14.8	3.8	22.7
Spruce, Engelmann[2]	64.0	5.2	25.9	10.9	22.2
Spruce, black	51.6	5.0	14.6	4.4	28.0
Pine, jack	62.6	8.0	12.4	3.0	41.3
Pine, lodgepole	72.9	28.7	10.9	5.6	29.8
Pine, slash	48.5	3.4	10.6	3.7	28.9
Pine, sugar[2]	62.7	1.5	21.7	3.2	36.0
Pine, western white	57.3	3.2	7.5	14.3	33.3
Hemlock, eastern[2]	51.7	2.8	21.2	3.3	24.6
Boxelder	39.7	2.4	6.3	6.2	23.7
Maple, sugar	28.3	1.2	3.9	2.4	19.2
Alder, red	37.8	2.3	3.9	3.7	27.5
Birch, yellow	46.9	4.3	10.8	2.3	28.4
Birch, paper	42.9	9.4	10.5	2.5	25.1
Pecan	50.9	0.8	18.4	5.4	25.3
Sweetgum	48.3	1.5	17.7	7.4	21.3
Blackgum	39.4	2.5	4.6	5.3	27.8
Sycamore, American	33.4	2.1	6.0	3.6	22.0
Cottonwood, swamp	35.0	1.9	8.0	4.8	20.2
Aspen, quaking	41.8	4.0	11.6	4.7	22.0
Oak, white	38.2	2.7	4.4	5.8	26.5
Oak, northern red	39.8	4.8	7.9	3.6	22.3
Willow, black	35.3	1.6	3.8	4.8	23.8
Elm, American	45.2	0.5	10.1	6.0	27.0

[1] An appreciable percentage of this extract was reprecipitated on acidification.
[2] Barks with highest tannin content.

deriving bio-chemicals from lodgepole pine bark. We discuss conversion techniques used on lodgepole pine bark including alkaline extraction, solvolytic liquefaction, and alkoxylation.

7.1.3 Aim of this article

Many studies have shown that bark has good potential to produce phenolic adhesives and polyurethane foams, since the components with an aromatic

Table 7.5: Chemical composition of barks [17].

Bark	Hex-soluble	Eth-soluble	1 % NaOH	Holocellulose	Lignin	FCP*1	SN*2	TP*3
Lodgepole pine	15	11.7	35.5	28.9	8.9	8.3	23.9	279
Aspen	8.6	22.3	26.2	30.2	12.7	3.7	14	253
White birch	1.9	3.1	23.7	44	27.3	6.8	28.9	297
Sugar maple	2.3	5.5	20.5	44.8	26.9	2.6	12.5	236
Balsam fir	8.9	4.5	32.6	37.9	16.1	12.5	38.4	131

*1 FCP = Formaldehyde-Condensable Polyphenols in 1 % NaOH soluble extractives
*2 SN = Stiasny number
*3 TP = Total phenolics of ethanol-soluble extractives

polyol structure in tree bark can partially or entirely substitute the phenol-like structures in petroleum-based resins and foam production. Among various bio-based thermosetting polymers, bark-based resins are considered close to market-readiness. The aim of this article is to summarize some insights regarding the bark liquefaction and extraction processes, and discuss the general properties of bark-based resins and foams. We hope this article can provide researchers with a comprehensive understanding of applying sustainable chemistry for bark utilization.

7.2 Bark conversion into chemical precursors

Because bark possesses a complex chemical composition, without a properly chosen conversion technology, the heterogeneity of bark may result in impaired properties and filler behaviors compared to adding a homogenous compound. Thus, the most fundamental step towards making valuable chemicals from bark is to extract or convert target compounds from the biomass. There are many ways to convert biomass into chemical precursors, including gasification, pyrolysis, enzymatic treatments, extraction, solvolytic liquefaction, and alkoxylation. However, gasification and pyrolysis techniques are the primary means for extracting low-molecular-weight products and gases. In gasification, bark, a diverse macromolecule of molecular greater than 100,000 Da, is converted into gases with molecular weights under 300 Da. This process requires high energy usage that raises the carbon footprint of the procedure. Because of their lower energy usage, extraction, solvolytic liquefaction, and alkoxylation options may be seen as more attractive choices for producing bark-derived adhesives and foams. These methods also have the benefit of being low cost and using recyclable solvents to reduce side reactions. In the following sections, we discuss three main types of conversion technologies for biomass: alkaline extraction, solvolytic liquefaction, and alkoxylation.

7.2.1 Alkaline extraction – NaOH extraction

Alkaline extraction does not require toxic organic solvents and can be conducted in mild conditions. In the past, people used alkaline solutions to extract natural plant dye from bark for clothes or leathers. Additionally, compared to other solvent extractions, alkaline extraction is the most common treatment for obtaining polyphenolics due to its high yield and low viscosity solutions. Among the common bases for biomass extraction (NaOH, Na_2SO_3, and Na_2CO_3), NaOH provides the highest extraction efficiency and the highest condensable polyphenol content [13]. Thus, this section focuses on the NaOH extraction.

Extraction in general is a solid-liquid leaching process. The leaching efficiency can be affected by many experimental parameters such as time, temperature, liquid to solid ratio, and solvent concentration. A previous study has shown the extraction process can be simplified and described by a modified extended Fick's law analytical solution model based on the total yield [14]. However, the leaching efficiency is far more complicated than what can be evaluated only by the total yield. It has been shown that a high concentration of sodium hydroxide can improve the total yield, while decreasing the reactivity of phenolic compounds due to the formation of phlobatannin, as shown in Figure 7.3 [15]. Thus, side reactions such as reorganization and re-condensation can significantly impact the results. To address this issue, researchers have found that limiting the NaOH concentration to 2.5 wt% or below and adding a nucleophilic agent like urea can reduce the aggregation [16].

The most common ways to measure the reactivity of phenolic compounds for adhesive application are by the Stiasny Number method (which reflects the reactivity of phenolic rings to formaldehyde), Folin–Ciocalteu method [17], and nuclear magnetic resonance (NMR) analysis [18], which detect the amount of total phenolics or the abundance of phenolic hydroxyl groups. Zhao et al. [19] showed that NMR analysis can offer detail on molecular structures of extractives. For example, compared to regular extracted compounds, autoclave-assisted alkaline extractives have fewer C_4 to C_8 and C_4 to C_6 interflavonoid bonds, more degraded lignin fragments, and more hemicellulose (as shown in Figure 7.4), resulting in higher molecular weights, higher viscosities, and shorter gel times of the derived phenol-formaldehyde resins [19].

Biomass conversion is a complex process. Even a mild alkaline extraction possesses complicated polymer degradation, modification, and condensation reactions. It is crucial to carefully control the extraction parameters to obtain the target compounds. However, extraction suffers from low yields overall, usually lower than 50 %. The following sections will focus on methods that attempt to increase the yield of extractions, including a relatively harsh depolymerization of the biomass (solvolytic liquefaction: phenolysis and alcoholysis) and the polymerization of the biomass (alkoxylation).

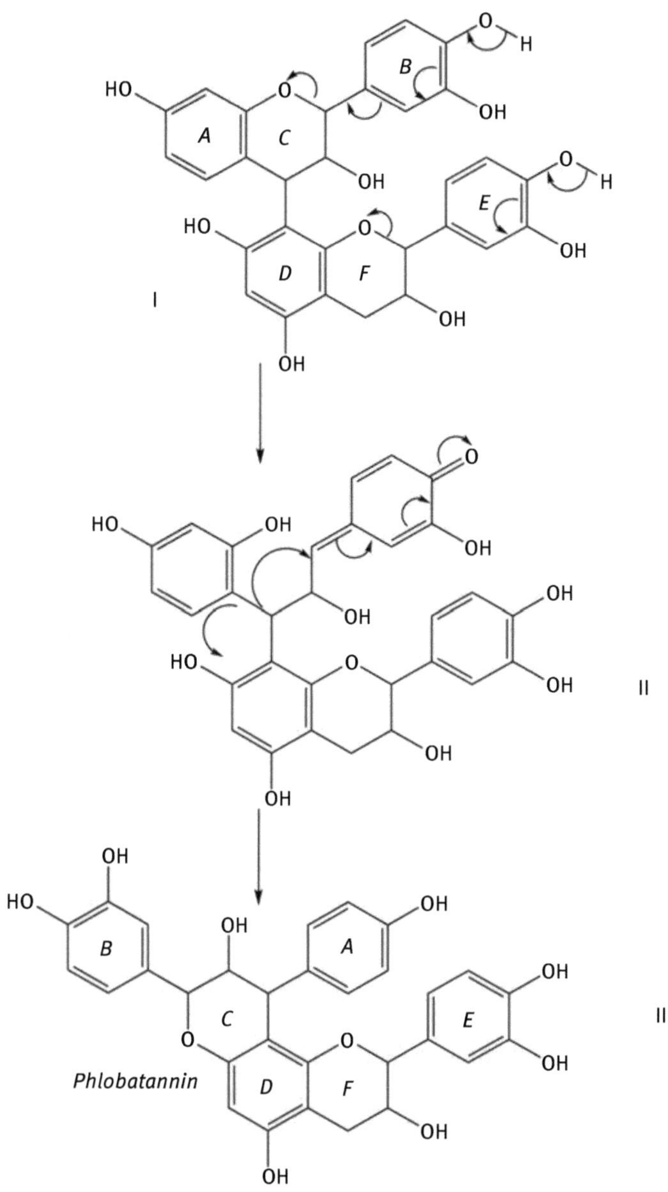

Figure 7.3: The formation and structure of phlobatannin [15].

7.2.2 Solvolytic liquefaction: phenolysis and alcoholysis

Similar to extraction, solvolytic liquefaction is also a degradative process, but solvolytic liquefaction generally requires higher reaction pressures (10–30 MPa) and

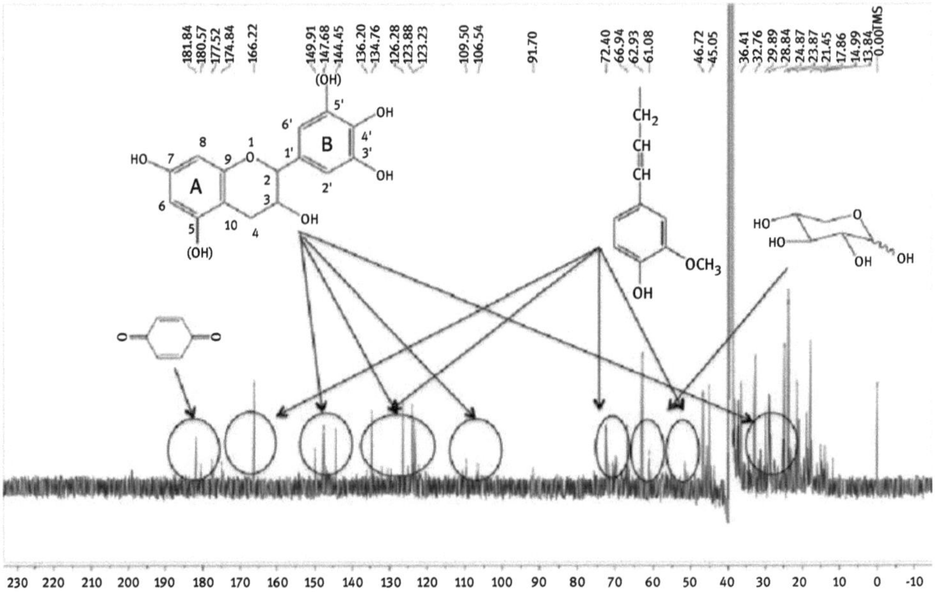

Figure 7.4: Structures identified in an NMR scan of autoclave-assisted alkaline extractives from lodgepole pine [19].

higher reaction temperatures 200–350 °C leading to relatively low residuals of insoluble solids compared to extraction. More recently, it has been shown that the liquefaction temperature can be decreased to 80–170 °C and conducted at relatively low pressure (<10 MPa) with the assistance of particular solvents and acid catalysts [20]. According to the types of solvent, the solvolytic liquefaction can be divided into two types, phenolysis and alcoholysis, which are detailed below.

7.2.2.1 Phenolysis

The phenolysis reaction is performed using phenol as a solvent and aims to extract the soluble fragments from bark. The reaction process and reaction parameters are shown in Figure 7.5. The reactivity of the targeted products is determined by their hydroxyl functionality. An ideal liquefaction process would minimize the amount of residuals of insoluble fragments and maximize the reactivity of extracted compounds. The use of phenol liquefaction on biomass conversion has been extensively studied and the liquefied products are generally applied in phenol-formaldehyde (PF), melamine-formaldehyde (MF), urea-formaldehyde (UF), and epoxy resins. The global phenolic resin market was valued at $10.66 billion in 2015 and is projected to reach $15.01 billion by 2021 [21]. Thus, the market size reflects a huge future demand of bio-derived phenols, which is more environmentally-friendly and less toxic, to replace petro-phenols.

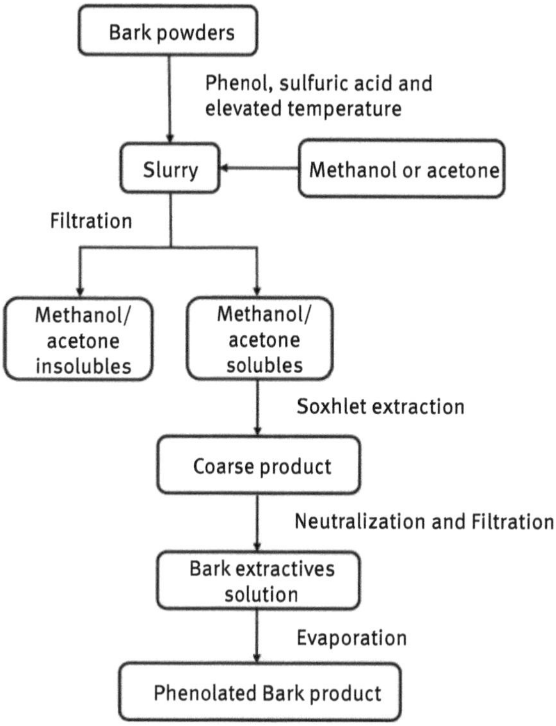

Figure 7.5: The general process of bark phenolysis.

Phenolysis liquefaction can be catalyzed by acids, bases, or metal salts. During the liquefaction process, some phenols condense their reactive sites to a stable ether bond, diminishing their activity toward formaldehyde leading to low reactivity and low yield. Among these three types of catalysts, acid catalyzed products retain the greatest reactivity owing to its lowest amount of condensed phenols at the lowest cost. An acid-catalyzed phenol liquefaction process can be roughly divided into three reactions as shown in Figure 7.6 [22, 23]. The earliest reaction shows lignin, hemicelluloses, and amorphous cellulose quickly being attacked by the acid. Some reports suggest lignin is the first component to be emitted from bark under the phenol liquefaction because of its susceptibility to phenol-like solvents [24]. During the second reaction, acid gradually penetrates into crystalline cellulose thereby dissolving it. Thus, cellulose is the last component to be removed from bark and has a decisive role in the speed of the liquefaction reaction. The third reaction is the re-condensation reaction, a side reaction occurring simultaneously, which is promoted by isolated lignin or phenolic extractives. During the entire process, it has been shown that insoluble components decreased at first because the bark degraded into small fragments but then increased over time at higher reaction temperatures, which indicated that these re-condensed compounds can precipitate out of solution.

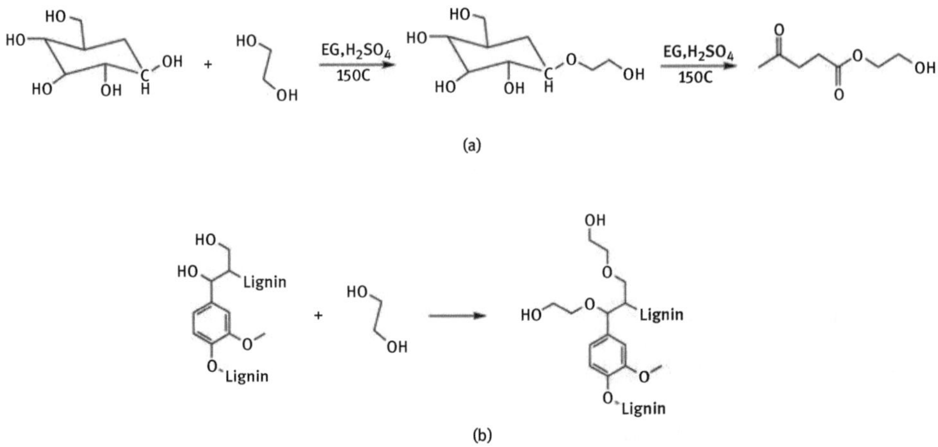

Figure 7.6: Degradation of single sugar to levulinate esters. Reprinted with permission from D'Souza J. The Utilization of Bark to Make Rigid Polyurethane Foams. Toronto: University of Toronto Thesis; 2016.

To address this issue, a high solvent to biomass ratio has been used to achieve high yields and prevent re-condensation reactions. Apart from re-condensation, another critical side reaction is sugar degradation. In a harsh environment, cellulose is degraded into levulinate esters (Figure 7.7) which reduced its hydroxyl functionality from three to one [16].

Other influential factors include the types of acids used, the catalyst loading, the reaction temperature, and the reaction time. Our previous results have shown that the liquefaction yield, free phenol content in the liquefied bark fraction, and properties of residuals were affected mainly by the catalyst loading, and to a lesser extent by the reaction temperature, reaction time, and phenol/biomass ratio [25]. We have found previously that optimal liquefaction conditions for lodgepole pine bark was at a temperature of 150 °C, reaction time for 120 min, phenol/bark mass ratio at 3, and H_2SO_4 content of 3 wt% [26]. Additionally, we have found that phenolated lignin products, such as phenylcoumaranes, benzocyclobutanes, triphenylethanes, diphenylmethanes, and guaiacol structures and fragments, as well as phenolated cellulose products, such as tri(4-hydroxylphenyl)methane and 1,1,2-tri(4-hydroxylphenyl)ethane, are found in liquefied bark [27].

Note that the residual phenols from liquefaction may not significantly impact the final properties of the phenolic resins, but they have a deleterious effect on the properties of polyurethane products made with these resins because the remaining phenols can terminate polymerization reactions. This fact reminds us that it is essential to carefully match the conversion technology with final applications of the bio-derived products.

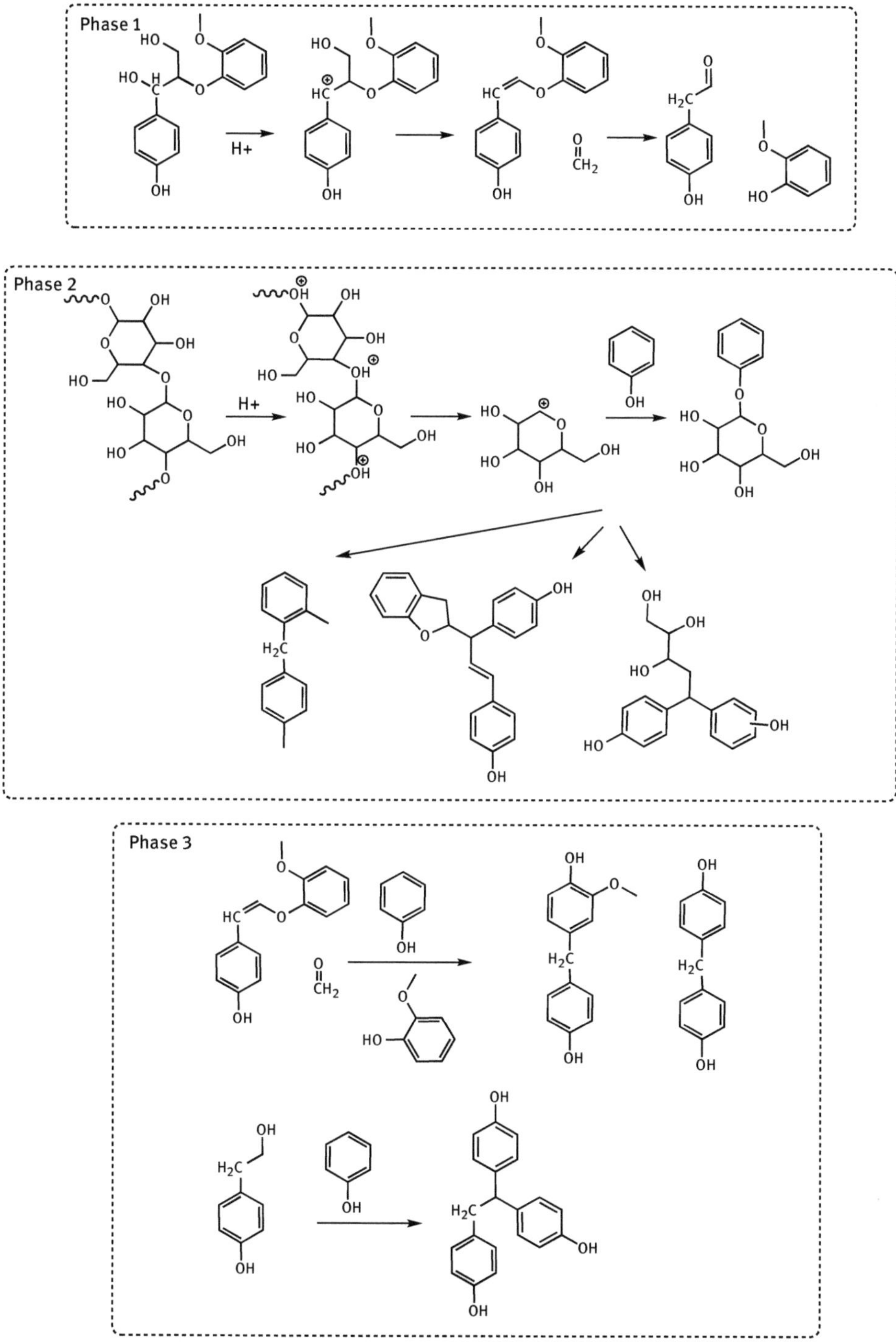

Figure 7.7: Three major reactions during bark phenolysis liquefaction [23].

7.2.2.2 Alcoholysis (polyhydric alcohols)

The alcoholysis liquefaction reaction uses polyhydric alcohols such as polyethylene glycol (PEG), polypropylene glycol (PPG), glycerol, and some low-boiling alcohols to dissolve the bark. The flowchart of alcoholysis liquefaction and its reaction parameters can be found in Figure 7.8. Alcoholysis is a more popular conversion technology than phenolysis for the polyol component in polyurethane due to the harmful effect of the phenols that remain after phenolysis. These two conversion technologies share many similarities, including being catalyzed by acid, a reaction mechanism that can be divided into three phases, and the high solvent/biomass ratio as a strategy to suppress side reactions and improve yield. Alcoholysis liquefaction has been less studied than phenol liquefaction, so influences of some synthesis conditions on final properties of bio-polyols are still uncertain.

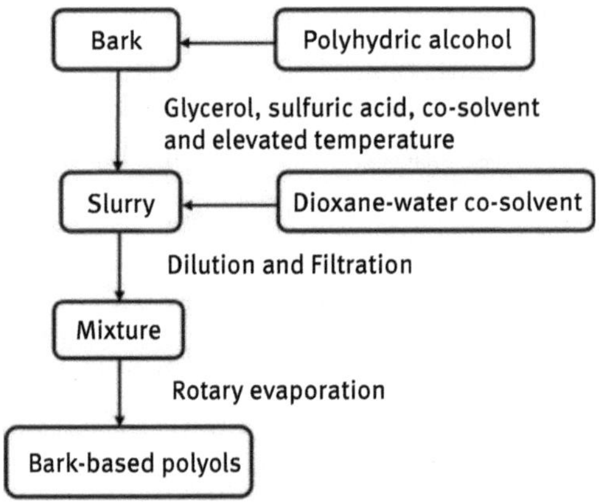

Figure 7.8: A general process of bark alcoholysis.

Generally, the reaction temperature of bark alcoholysis ranges from 90 to 250 °C and many studies have shown that the temperature has a significant influence on yield, degradation, and compositions. D'Souza and Yan [28] reported that synthesis at high liquefaction temperature (160 °C) can efficiently improve the total yield of products and decrease the undissolved residuals compared to that of low liquefaction temperature (90 °C). However, the holocellulose degrades to levulinate and formic acid and thus their derived products contain many components with low hydroxyl functionality. In addition, at high liquefaction temperature, more lignin was extracted out from bark leading to a higher amount of aromatic components in final products; these lignins may undergo a re-condensation reaction with a consequent low yield. There were no phenolic structures observed using FTIR, ^{13}C-NMR, and ^{31}P-NMR methods at various liquefaction temperature (90 °C, 130 °C, and 160 °C), which suggests that the phenolics were converted to aliphatic alcohols through condensation reactions with polyhydric alcohols. It has been shown that liquefaction at 130 °C is ideal for producing a polyol with a relatively broad molecular weight distribution, whereas liquefaction at 90 °C or 160 °C produced bio-polyols with a large amount of low molecular weight compounds.

The selection of polyhydric alcohols and organic co-solvents can also be quite impactful on the liquefaction yield and the polyol characteristics. Liquefaction yield is highly tunable with the use of polyhydric alcohols with primary hydroxyl groups. The low equivalent weight alcohols can provide the highest liquefaction yield up to 59.3 % [29]. It has been shown that highly polar hydroxyls (primary) and short chains created a highly protic solvent that improved conversion and protected the biopolymers from degradation. The ^{1}H NMR analysis (Figure 7.9) indicated that a greater amount of sugar degradation products were observed when polyhydric alcohols with secondary hydroxyl groups were used. Regarding organic solvents, ketonic solvents showed the greatest increase in the liquefaction yield. There is evidence that the highly polar carbonyl group of ketonic solvents like acetyl acetone and cyclohexanone may hinder condensation side reactions [29].

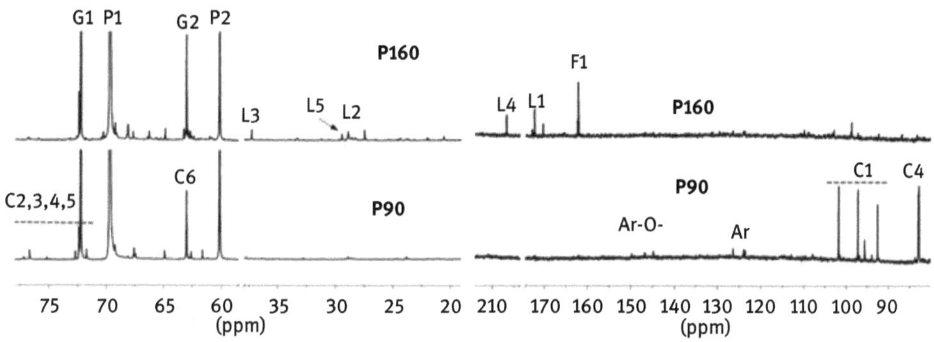

Figure 7.9: Degraded sugars were detected by ^{13}C-NMR (L and F mean levulinate and formic acid, respectively) [29].

Other important factors in alcoholysis are time, catalyst content, and co-solvents. These factors are not fully independent from temperature and polyhydric alcohol structures. Therefore, more systematic research is required to explore the effects of synthesis conditions on bio-polyols. With a better understanding of how synthesis conditions and solvent impact liquefaction, it will be beneficial for the utilization of the bio-polyols for improving the viability of polyurethane foam products.

7.2.3 Alkoxylation: Oxypropylation

The definition of alkoxylation is to add an epoxide to a hydroxyl-containing compound such as glycerol or sugars. Alkoxylation is a common method to produce polyether polyols in industry because alkoxylation reactions can control the extended chain length (from propylene oxide) of the seed compounds (-OH compounds), which results in a product with tunable properties. Here we will describe the manifestation of alkoxylation as oxypropylation using propylene oxide (PO) to produce chain-extended bio-polyols.

There are three major advantages of using oxypropylation on producing bio-polyols: (1) the hydroxyl groups, particularly the phenolic ones that are entrapped inside the molecule, are liberated from steric and/or electronic constraints; (2) the process can be done under alkaline conditions that will lessen polysaccharide degradation and maintain the hydroxyl functionality of seed compounds; (3) oxypropylation is a mature commercial process in industry to produce polyols from seed compounds, so the technical barriers for industry to produce bio-polyols will be lower. The process of alkoxylation can be divided into the following four phases (Figure 7.10) [16]:

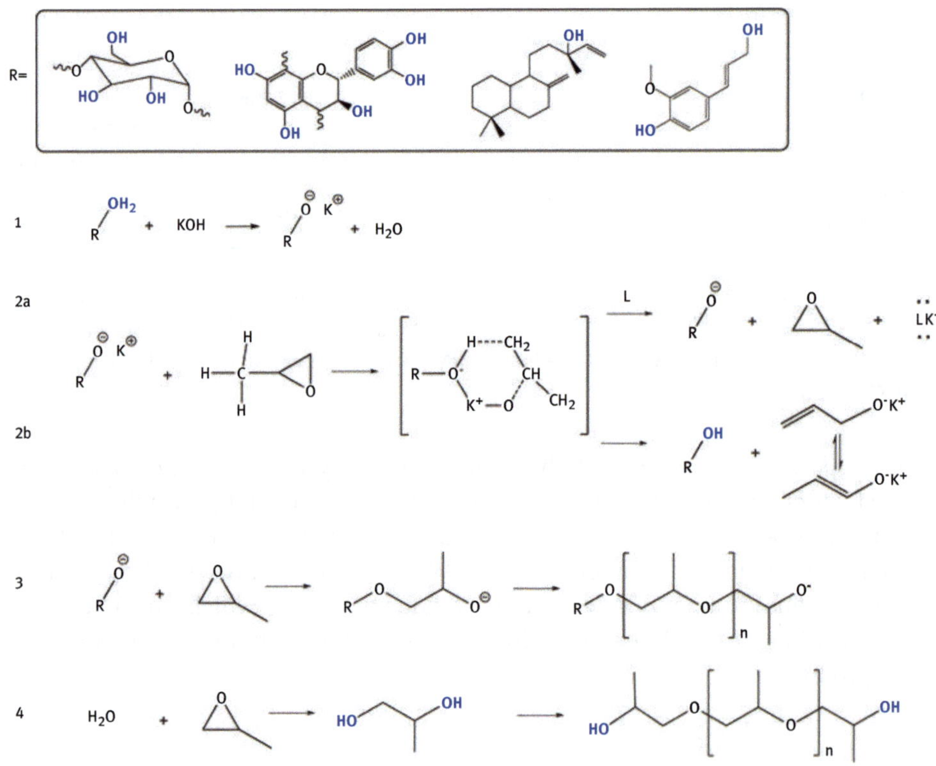

Figure 7.10: General reactions of oxypropylation on bark components. Reprinted with permission from D'Souza J. The Utilization of Bark to Make Rigid Polyurethane Foams. Toronto: University of Toronto Thesis; 2016.

As shown in Figure 7.10, the first reaction is that KOH attacks on the hydroxyl groups of seed compounds and then produces water and alcoholate which promotes the following chain propagation reaction. The second reaction shows chain propagation between seed compounds and PO to form a copolymer, which is the major products after an oxypropylation reaction. The third reaction shows a propagation between water and PO to form a homopolymer which is the secondly main product of oxypropylation. However, there are some by-produces which occur under certain condition as shown in reaction four. Through a chain rearrangement, the alcoholate turns back to alcohol-seed compounds and open the ring structure of PO to form an allyl alcohol which terminates the chain propagation and forms small undesired products. Generally the side reaction is negligible, but it can become pronounced with high PO/seed compound ratio.

As shown in the mechanism, the major side reaction is the formation of unsaturation on PO, which is used as an intentional reaction for synthesizing high molecular weight polyols at high PO/seed ratio. This side reaction suggests that an appropriate base catalyst can have a significant influence on the properties of products and the cost of material waste. KOH is an inexpensive catalyst suitable for industrial scale but consumes a certain amount of PO during side reactions. Thus, double metal cyanide (DMC) catalyst can be a better option to reduce the formation of unsaturation with very limited usage (0.015–0.02 meq per g) [30], but the price of DMC is much higher than KOH. Thus, the appropriate selection of the catalyst for biomass conversion is crucial to design if large molecular weight polyols (>1000 Da) are targeted.

Apart from the catalyst, the molecular structure of seed compounds can also possess significant effects. Usually, the presence of acid groups on seed compounds can slow down the oxypropylation [31, 32], and previous research shows sodium ions can stabilize alkali lignin, which displays higher reactivity than organosolv lignin, a lignin separated from lignocellulostic material by solubilization in organic solvents [33]. However, our lab results show oxypropylated bark (79 %) has a higher conversion rate than the oxypropylated alkaline extract of bark (32 %) at the same reaction time and temperature [34]. This suggests that the concentration of phenolics and acid in the alkaline extract can significantly retard the reaction rate despite the polyphenols being stabilized by sodium ions. In the same study, we concluded that oxypropylation might be a more suitable conversion technology to prepare bio-polyols than liquefaction for the following reasons: (1) oxypropylated bio-polyols contain roughly 80 % secondary alcohols, 20 % primary alcohols, and no phenolic alcohols, so that no unstable urethane linkage will be formed by phenolic alcohols, which allow secondary alcohols to react with isocyanate and form polyurethane foam cells in a reasonable period; and (2) oxypropylated bio-polyols have a longer shelf-life compared to liquefied bio-polyols, which tend to form sediment over time.

Compared to liquefaction, the studies of alkoxylation on biomass conversion are still relatively few. Also, the reported synthesis conditions of alkoxylation are varied with diverse range of pressures, solvents, reaction times, and biomass types [35–39].

More research on synthesis conditions of alkoxylation are required in order to successfully apply this technique to biomass conversion.

7.3 Bark-derived adhesives and foams

7.3.1 Bark-derived adhesives

Due to the high amount of polyphenols in bark, the usage of bark powder [40, 41], tannins [42, 43], and extracted bark components [44, 45] in producing phenol-based resins extends back to the mid-twentieth century. At the time, many researchers focused on using bark components for wood panel applications such as plywood [46] and particle board [47]. A previous review paper [48] has summarized these applications. In this article, we will concentrate on the use of alkaline extractives and liquefied bark in the production of phenol-formaldehyde (PF), melamine-formaldehyde (MF), and epoxy resins. For each resin, we will discuss the effect of introducing bark components on resin structures, mechanical/thermal properties, curing performance, and summarize the suitable synthesis conditions.

7.3.1.1 Phenol-formaldehyde resins (PF)

PF was the first commercial synthetic resin and has been widely used in many applications such as coatings, thermal insulation, adhesives, and printed circuit boards. Due to its excellent water resistance, superior mechanical properties, and strong adhesive bonds on substrates, the usage of PF resins is expected to grow 6.7 % per year from 2016 to 2022 to reach a value of $16.05 billion [49]. PF resin can be divided into two classes: one is resole, a thermosetting resin, made with a formaldehyde to phenol ratio of greater than one and catalyzed by a base solution; the other one is novolac, a thermoplastic resin, made with a formaldehyde to phenol ratio of less than one and catalyzed by an acid solution. Many studies have reported successful replacement of petroleum phenols with natural bark-derived polyphenols in both types of resins. Generally, bark-derived PF resins have higher molecular weights, higher polydispersity indices, shorter gel times, and faster curing rates than the commercial PF resins [26]. These properties will be explained in detail in the following sections.

7.3.1.1.1 Bark-derived resole resins

According to the starting materials, bark-derived resole resins can be divided into two main categories: those with alkaline bark extractive components (abbreviated as AE) and those with liquefied bark components (abbreviated as LB). More results of bark-pyrolysis oil-based resole resins can be found in other review papers [48].

7.3.1.1.2 Derived from alkaline extractive components (AE-PF resole resin)

Compared to the increase in demand for tannins and lignin used in adhesives, the increase in usage of alkaline bark extractives as adhesives instead has been far slower. In the late twentieth century, the function of alkaline bark extractives in PF resins was just for accelerating curing and reducing the viscosity, since they have been shown to minimize gelation and shorten press time [50, 51]. In the last decade researchers have successfully synthesized resole PF resins using alkaline extractives from bark without chemical pre-treatment [52].

Extractives at 30 wt% replacement of phenol exhibited similar dry and wet bond strengths to commercial PF resins. At a 50 wt% substitution level, AE-PF resins had dry and wet bond strength similar to the commercial PF resin. Compared to PF resin made from liquefied bark, AE-PF resins have better thermal resistance due to differences in chemical composition and resin structures [53]. As described above, alkaline extractives contain a high number of aromatic structures from polyphenols and lignin resulting in high thermal resistance. In addition, the bark extractives made the ortho position of phenol react more favorably with formaldehyde than the para position [27]. Linkages made through the ortho-position are more stable than those through the para linkages; furthermore, ortho-ortho or ortho-para linkages have much higher adhesion to cellulose substrates than para-para linkages [54]. However, alkaline extractives have lower reactivity toward formaldehyde than liquefied bark.

To address this issue, the same research team prepared autoclaved bark extractives to obtain higher amounts of tannins in the extractives, which lead to a higher Stiasny number (see section 7.2.1) [19]. Compared to the regular bark extractive PF resins, autoclave bark extractive PF resins have lower viscosities and similar mechanical/thermal properties.

7.3.1.1.3 Derived from phenolized bark (LB-PF resole resin)

Phenol not only dissolves the main components of the bark but also reacts with depolymerized lignin and cellulose. The phenolated products are very reactive with formaldehyde, which makes them an ideal material to partly replace petroleum-based phenol with which to synthesize LB-PF resins.

The development of LB-PF resins occurred much earlier than AE-PF resins resulting in a more thorough understanding of the relationship between reaction conditions and the properties of LB-PF resins. For the catalyst, NaOH is the most common base to condense phenols and formaldehydes. The level of NaOH added has an important effect on various properties of bark-based phenol formaldehyde adhesives, such as gel time, free formaldehyde content, thermosetting peak temperature, molecular weight distribution, wet shear strength, and free formaldehyde release from the bonded plywood. A two-step process of adding NaOH was found to be a better method in preparing bark-based PF adhesives [55]. It is worth noting that sufficient NaOH is needed to neutralize the acidic phenolated bark first and then to catalyze the resole synthesis.

Phenolized bark from different bark species and liquefied with different catalysts can also produce different properties in PF resins. For the liquefied catalyst, it is well-known that phosphoric and sulfuric acids produce a lower residue ratio after phenolysis than hydrochloric acid or oxalic acid [56], but resole resins prepared from bark liquefied with sulfuric acids had higher viscosity. Bark species can affect the resole properties, e. g., phenolated Chinese fir bark–based resole displayed higher viscosity (840 mPas) than Taiwanese acacia bark–based resole (350 mPas) [57]. Zhao et al. showed that beetle infestation has no negative effect on the bonding properties of the bark-PF resin [25].

Compared to liquefied tannin PF resins, LB-PF gives better wood failure and shear strength results [58]. Tannin modification, such as sulfitation, acid hydrolysis, or alkaline hydrolysis to break down the hydrocolloid gums in tannin extractives, can also be applied to improve the resulting resin properties. Fortification or copolymerization with PF is found to be another practical approach to enhance resin performance.

7.3.1.1.4 Bark-derived novolac resins (LB-PF novolac resin)

In contrast to the one-stage reaction to produce resole resin, novolac resin is a partially cross-linked linear thermoplastic polymer which involves an additional synthetic step to add the curing agent to achieve a fully cross-linked structure. Cured novolac resins can be completely different from resole resins depending on the curing agents. The synthesis processes of both resole and novolac resins are summarized and compared in Figure 7.11.

The following discussion will be divided into two parts. The first part describes uncured LB-PF novolac and the second part describes cured LB-PF novolac resins. The reported uncured LB-PF novolac oligomers are in a solid form, and the molecular weight of synthesized resins increased as the formaldehyde to phenol molar ratio (F/P ratio) increased in the resin formulation. Due to the increased molecular weight and reduced reaction sites on phenols, the onset curing temperature of LB-PF novolac increased as F/P ratio increased. The catalyst type also has been shown to have a significant influence on molecular weight and thermal stabilities. Sulfuric acid–catalyzed LB-PF novolac oligomers have higher molecular weights than HCl-catalyzed LB-PF novolac oligomers that increased the onset curing temperature and improved their thermal stability [59]. Unlike the lab-prepared neat PF oligomers, both LB-PF novolacs show a single exothermal peak instead of two.

Hexamethylenetetramine (HMTA) is the most common curing agent for novolac resins. It decomposes and reacts with phenolic rings in the resin during the initial stage of the curing process to form several intermediate products such as benzylamine and benzoxazine. These intermediate products go through further condensation

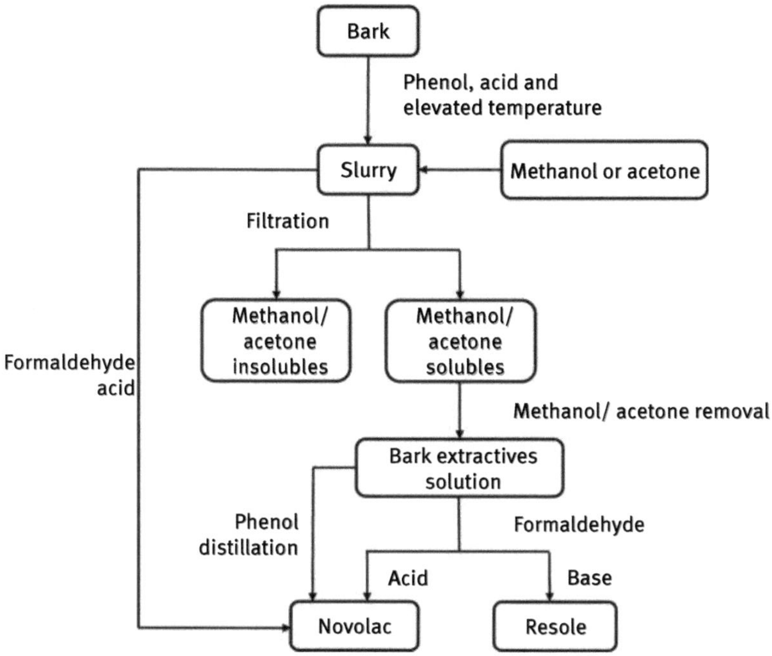

Figure 7.11: Comparison of manufacturing processes between novolac and resole resins [48].

reactions and eventually form macromolecules [60]. However, the same authors failed to prepare cured sulfuric acid-catalyzed LB-PF novolac resins due to the incorrect ratio between curing agent and novolac resins [60]. In addition, catalyst loading was shown to have a significant impact on the tensile properties. As the catalyst loading increased from 0.5 wt% to 2 wt%, their tensile strength improved for all the bark species, especially for LB-PF derived from pine bark, whose strength rose to 45.2 MPa. The results of modulus and elongation were affected more by the bark species than catalyst choice [61].

Overall, the bio-based PF resins are promising, environmentally friendly and lower cost alternatives to PF resins formulated solely from fossil fuel derived phenol.

7.3.1.2 Melamine-formaldehyde resins

Melamine-formaldehyde (MF) resin is the third most popular phenolic type resin in the global market. Compared to urea formaldehyde (UF) resins, MF has higher moisture-resistance and is mainly applied in exterior wood products. Bark extractives obtained from the mountain pine beetle-infested lodgepole pine showed promise as a suitable partial replacement for melamine in MF resin formulations, and the possible structure of bark extractive-MF resin is shown in Figure 7.12 [62]. Bark alkaline extractives from the mountain pine beetle infested lodgepole pine were used to replace

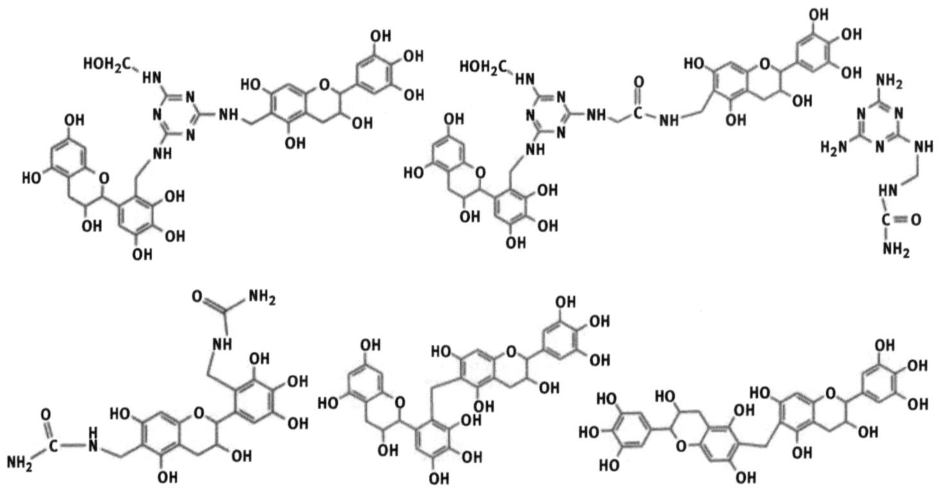

Figure 7.12: The proposed structure of bark extractive-based MF resins [62].

30 wt% of melamine in the formulation of bio-based bark extractive- MF resin. Results showed that the addition of the bark extractives used for resin formulation significantly increased the initial molecular weight, shortened the shelf life (with methanol as solvent), and showed competitive thermal stability and mechanical performance compared to the lab synthesized MF resins without any bark components inside. Further investigations of bark-based MF or MF/UF are required for understanding the effect of bark components on the chemical and physical properties of MF-based resins in the future.

7.3.1.3 Epoxy resins

Epoxy resins are thermosetting polymers that are widely used in many high-end applications such as aerospace and electrical insulation. The use of epoxy can also be very common in daily life such as dental fillers or fiber reinforced composites because of their many useful properties, such as the outstanding mechanical performance, strong chemical resistance, and superior dimensional stability. Nowadays, almost 90 % of the world production of epoxy resins is based on the reaction between bisphenol A and epichlorohydrin, yielding the diglycidyl ether of bisphenol A. However, with the recent health concern of bisphenol A and the impending shortage of fossil resources, researchers have begun to focus on developing alternative safe and green raw materials for resin synthesis.

Based on the analysis and experiments conducted by Kuo et al. [63], bark extractives show great promise to replace toxic bisphenol A and develop a new type of bio-based adhesive. Bark extractives were epoxidized after reaction with epichlorohydrin via a conventional epoxy resin synthesis. Successful functionalization of the bark

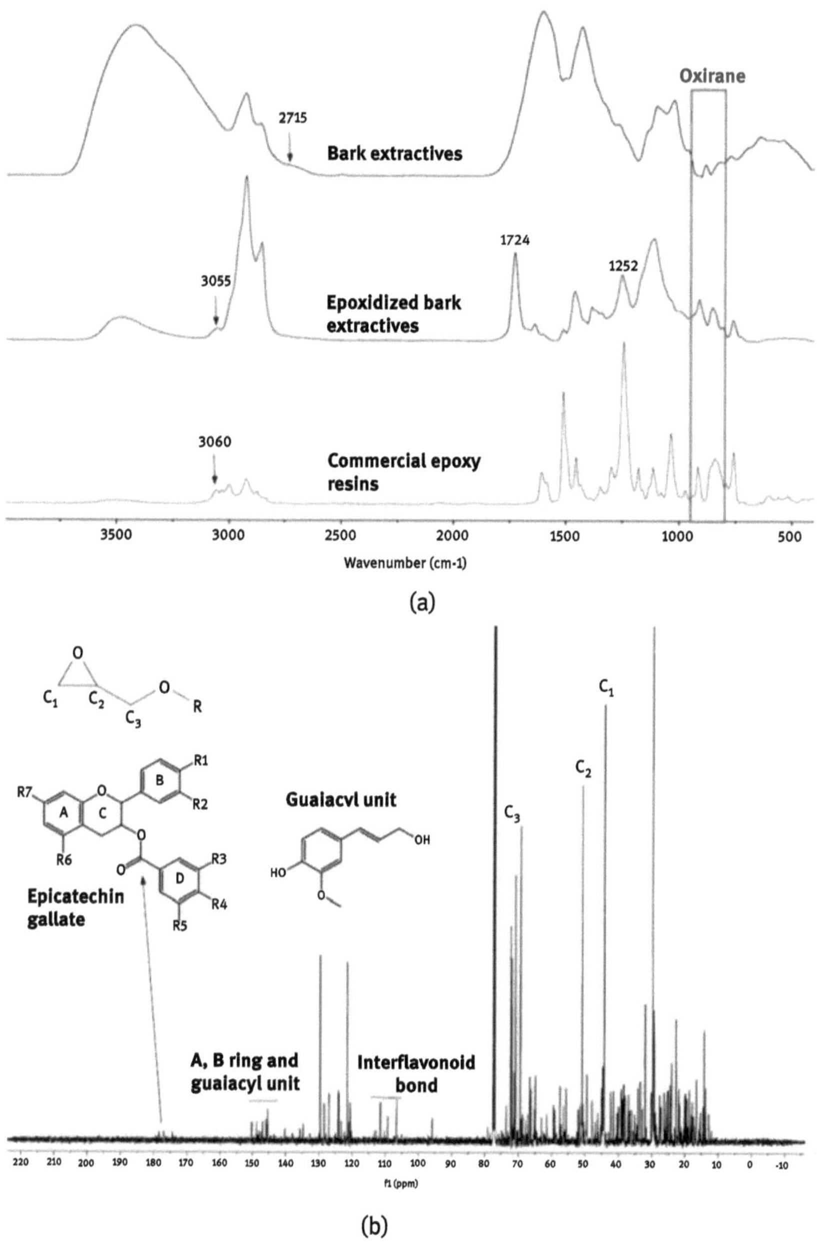

Figure 7.13: The results of FTIR and C-NMR to support the successful epoxidation reaction between bark and epichlorohydrin [63].

extractives was supported by the observation using FTIR and NMR (Figure 7.13). Bark extractive-based epoxy resins have a lower molecular weight compared to other polyphenol-derived epoxy resins like lignins, and the activation energy of the extractive-based epoxy resin is lower than that of the commercial products due to the abundant hydroxyl groups in extractives promoting the autocatalytic network formation. The bark-based epoxy resins can partially replace commercial epoxy resins up to 50 wt% and still have competitive mechanical properties. The thermal properties and the thermal degradation mechanism of the cured bark extractive-based epoxy resins have been discussed in detail, and based on the pyrolysis-GC-MS analysis the major components of bark-based epoxy resins were identified as methyl abieta-8,11,13-trien-18-oate, diethyl phthalate, 2,2-isopropylidenebis (3,5 -dimethylbenzofuran), and epimanool [64]

The reaction parameters have significant impact on the yield and reactivity of bark extractive-based epoxy resins. Researchers maximized the yield and epoxy content of the extractive-based epoxy resins by examining the glycidylation reaction with various substrates, solvents, catalysts, time periods, reaction temperatures, and sodium hydroxide/hydroxyl (NaOH/OHV) ratios. Spray-dried bark extractives were used as substrates due to their higher hydroxyl group content and lower molecular weight compared to the oven-dried bark extractives. A dioxane/water combination was selected from among four solvents based on the yield and epoxy equivalent weights of the final product, and tetrabutylammonium hydroxide (TBAH) was chosen as a ring-opening catalyst due to its effect of suppressing hydrolysis. The mechanism of glycidyl etherification catalyzed by NaOH and TBAH are both shown in Figure 7.14.

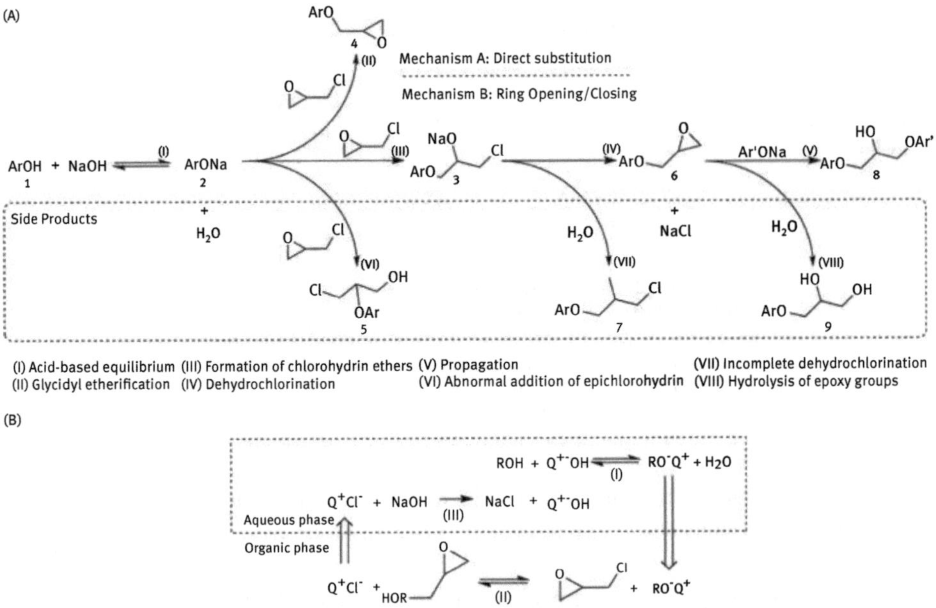

Figure 7.14: The reaction mechanisms between bark and epichlorohydrin (a) catalyzed conventionally using NaOH (b) catalyzed by phase transfer catalysis [65].

The maximum extent of conversion with minimum epoxy equivalent weight was achieved after 4.5 h with an NaOH/OHV ratio of 3.4 at 80 °C [65].

7.3.2 Bark-derived foams

7.3.2.1 Polyurethane foams (PUs)

Polyurethanes (PUs) are synthesized through an addition reaction of an isocyanate with a polyol. They are versatile engineering materials, which can be tailored by chemistry and produce diverse products including elastomers, foams, adhesives, sealants, and coatings. PUs are typically produced with petroleum-based materials, but the increasing environmental concerns have motivated researchers to replace polyols with sustainable resources such as bark, which is rich in polyols, as a viable replacement in PUs.

The characteristics of the polyols including molecular structure, functionality, molecular weight, and hydroxyl value and type have a strong impact on the final properties of PU foams. Synthetic polyols typically have a predicable structure in contrast to bio-based polyols that have a wide range of molecular weights. Bio-based polyols typically have hydroxyl values in the range around 300–600 $mgKOHg^{-1}$ for rigid foams, and a viscosity that can be easily tuned. Bio-based polyols with unknown functionality and a large range of molecular weights remain challenging for practical implementation [66].

The structure of liquefied bark typically used in PU applications is dependent on the parameters used during liquefaction. For instance, the solvent choice can affect the foam properties. As mentioned earlier in the section, phenolysis is not suitable for polyurethane applications because the remaining phenols in the solvent can terminate the polymerization. Alcoholysis liquefaction is a more suitable process but the parameters of the liquefaction can affect the final properties of the polyurethane. The most commonly used solvent in alcoholysis liquefaction is polyethylene glycol (PEG). PEG is also typically mixed with glycerol (PEG/G) to reduce the recondensation reaction of the phenolics without reducing the hydroxyl value. The PEG/G solvent resulted in reduced residuals that contained less lignin and more cellulose. As a result of these component differences, the foam made from PEG/G had a lower normalized compression modulus and higher density [67].

Temperature of the liquefaction reactions can also play an important role in the structure of the resulting polyols and final properties of polyurethane foams, as demonstrated by D'Souza et al. [28, 68]. Bark liquefied at lower temperatures (90 °C) contains more sugars and more lower molecular weight compounds. The low molecular weight moieties hinder the formation of a polymer network and do not allow the polymer chains to become fully incorporated in the polyurethane network. These low molecular weight moieties also contribute to the low functionality of the resulting foams and lead to decreased amounts of closed cells through possible antifoaming or plasticizing effects. The lower-temperature liquefied bark decreases modulus values and compression strength, when normalized by density. As shown in Figure 7.15, low temperature liquefied bark foams have a less defined morphology compared to the

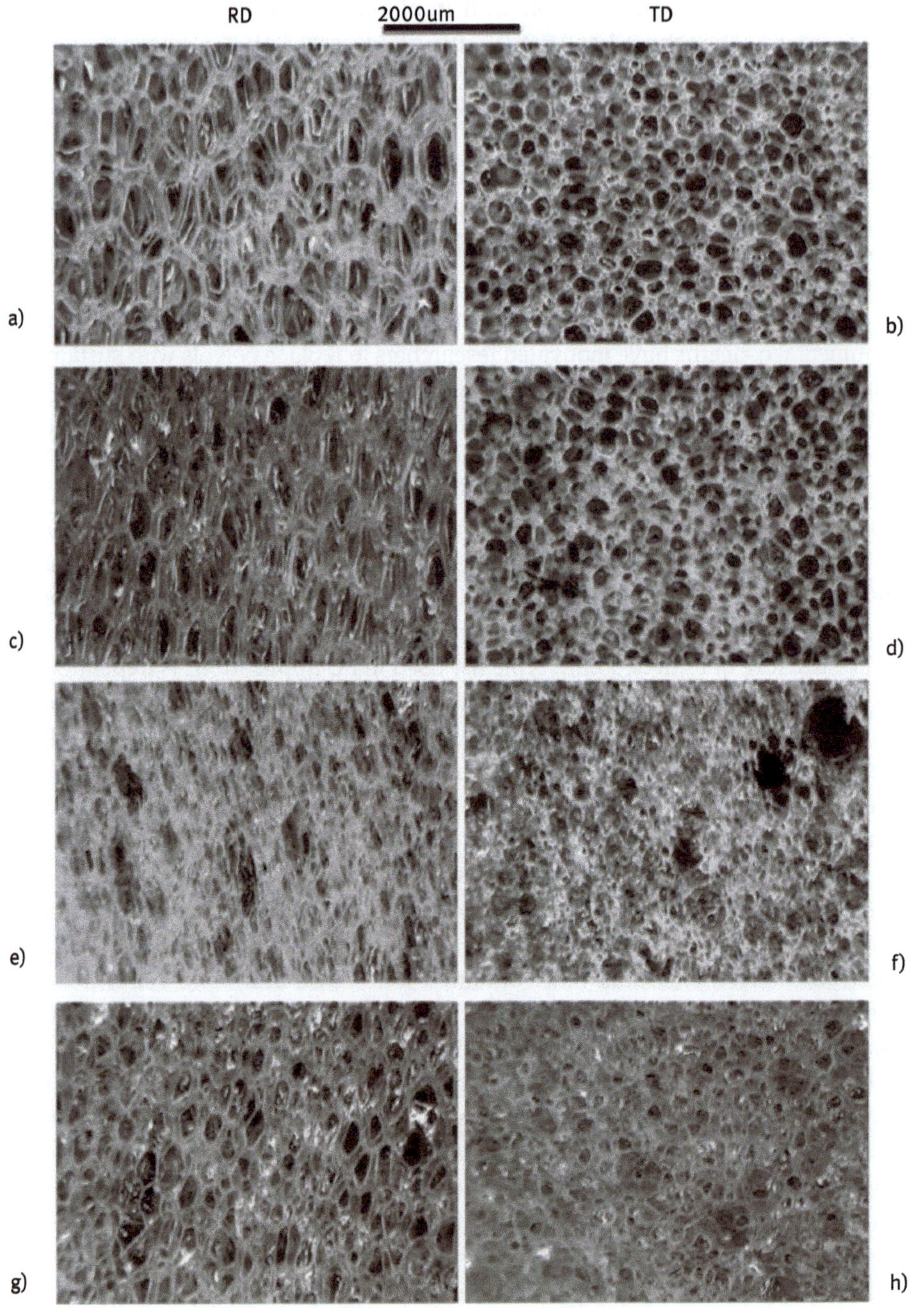

Figure 7.15: Optical micrographs of the foams based on (a,b) PPG-G; (c,d) PEG-G; (e,f) Liquefied bark at 90 °C; (g,h) Liquefied bark at 130 °C; in the rise (left images) and transverse (right images) directions [68].

controls and to the higher temperature liquefaction. It was speculated that the higher liquefaction temperature created large condensation polymers that aided in the formation of a stronger polymer network.

Grafting polymer chains on bark hydroxyl groups through oxypropylation addresses the issue of lower molecular weight components in the liquefied bark that reduces the quality of the polyurethane foam. Oxypropylation with polypropylene glycol takes place in mild alkaline conditions and reduces the degradation of the sugars. Due to the addition of polypropylene glycol, the number of primary alcohols is significantly increased. Primary alcohols react faster with isocyanates and create more stable urethane networking bonds compared to phenolic alcohols. For these reasons, in addition to the higher molecular weight, the bark that has gone through oxypropylation creates a superior network and improved mechanical properties [34].

Liquefied bark–based polyol showed good promise as a potential alternative to conventional petroleum-based polyols. Polyurethane adhesives containing liquefied bark had low gel times and good physical and mechanical properties. Furthermore, it was found that metal salts present in the liquefied bark can catalyze the reaction between isocyanate and polyol, further decreasing reaction times [69].

7.4 Summary and outlook

In the past decade, the development of bark-derived adhesives and foams has been intensively studied in a systematic manner due to the attractive features of bark including its sustainability, market readiness, and easy conversion. This article reviewed the chemical compositions of bark, the conversion technologies used to process bark, and the types of bark-derived adhesives and foams. There are three major conversion technologies summarized here: NaOH extraction, solvolytic liquefactions, and oxypropylation. NaOH extraction is a relatively simple and fast method, but its yield and the complexity of its components strongly depend on tree species. Solvolytic liquefaction can improve the yield and reactivity under more harsh reaction conditions, but degradation can occur during the process that harms the final properties. In addition, a large amount of chemicals and energy are consumed during solvolytic liquefaction. The third conversion technology is oxypropylation, a conventional way to extend polyol chains in the PU industry, and it shows strong potential to convert bark into bio-polyols. However, oxypropylation is mainly effective with aliphatic polyols, since the features of polyphenols might be degraded during processing. Thus, each conversion technology has its advantages and disadvantages depending on the required final application.

In the last section of this article, we summarized some recent findings on developing bark-derived adhesives and foams, including PF resin, MF resin, epoxy resin, and PU foams. All of them show competitive mechanical and thermal properties compared to the commercial products. However, there are three suggested areas that need future work. The first is to conduct a cradle to cradle life cycle analysis on bark-based polymers to access the sustainability, toxicity, and energy consumption of the polymers to find areas that can be improved. The second is conduct more comprehensive work using well-defined model compounds similar to bark components to better understand the effect of molecular structures on the performance of bio-based adhesives and foams. The last is to develop a thorough and complete characterization system to better evaluate the reactivity and general properties of bark converted uncured products. Our group has developed a set of characterization techniques on bio-polyols, but more studies are required for PF/MF/epoxy resins. It is our believe that further development of bark-based polymers will contribute to the production of high-value chemicals from bark using green chemistry for higher sustainability.

Acknowledgements: The authors would like to thank ORF-RE (Bark Biorefinery) and NSERC Discovery and DAS programs for the financial support and FP-Innovations as well as Huntsman for the technical support and guidance.

References

[1] Shmulsky R, Jones PD, eds. Forest products and wood science. 6th ed. Wiley-Blackwell. Chichester, UK; 2011.
[2] Natural Resources Canada CFS. The State of Canada's Forests In. Ottawa: Canadian Forest Service Publications, 2016: 72.
[3] Sandved KB, Prance GT, Prance AE. Bark: the formation, characteristics, and uses of bark around the world. Portland: Timber Press; 1993.
[4] Herrick FW. Useful products derived from conifer tree bark by chemical extraction. In: Acv V, eds. Converting bark into opportunities. Corvallis: Oregon State University; 1971.
[5] Graham AG. Hydroxyesters of bark phenolic acids. In: Google Patents; 1969.
[6] Allan GG. Reaction products of lignin and bark extracts and process for same. In: Google Patents; 1969.
[7] Bock LH, Herrick FW. Alkaline bark extraction and product. In: Google Patents; 1958.
[8] Allan GG. Esters of bark phenolic acids. In: Google Patents; 1970.
[9] Anderson JK, Herrick FW, Russell GK. Chemical products from bark. In: Google Patents; 1958.
[10] Pauley RD. Bark components as resin ingredients. In: Google Patents; 1956.
[11] Harkin JM, Rowe JW. Bare And Its Possible Uses. Bark and its possible uses. In: Service UF, ed. Madison, U.S. Department of agriculture 1971: 60.
[12] Patil ND, Yan N. Investigate cleavage of beta-O-4 linkage in lignin model compounds by aerobic oxidation of C alpha and C gamma hydroxyl groups. Tetrahedron Lett. 2016;57:3024–28.
[13] Vazquez G, Gonzalez-Alvarez J, Santos J, Freire MS, Antorrena G. Evaluation of potential applications for chestnut (Castanea sativa) shell and eucalyptus (Eucalyptus globulus) bark extracts. Ind Crops Prod. 2009;29:364–70.
[14] Wright GD. Mathematical modeling of the solid-liquid extraction of phenolic-rich compounds from pinus contorta bark. Toronto: University of Toronto; 2015.
[15] Sealyfisher VJ, Pizzi A. Increased pine tannins extraction and wood adhesives development by phlobaphenes minimization. Holz Roh Werkst. 1992;50:212–20.
[16] D'Souza J. The utilization of bark to make rigid polyurethane foams. Toronto: University of Toronto; 2016.

[17] Huang Z, Yan N. Characterization of major components in barks from five Canadian tree species. Wood Fiber Sci. 2014;46:167–74.
[18] Jiang ZH, Argyropoulos DS, Granata A. Correlation-analysis of P-31 Nmr chemical-shifts with substituent effects of phenols. Magn Reson Chem. 1995;33:375–82.
[19] Zhao Y, Yan N, Feng M. Synthesis and characterization of bio-based phenol-formaldehyde resol resins from bark autoclave extractives. Forest Prod J. 2016;66:18–28.
[20] Hu SJ, Luo XL, Li YB. Polyols and polyurethanes from the liquefaction of lignocellulosic biomass. Chemsuschem. 2014;7:66–72.
[21] Phenolic Resin Market by Type (Resol Resin, Novolac Resin), Application (Wood Adhesives, Molding, Insulation), End-use Industry (Automotive, building & construction, Furniture) – Global Forecasts to 2021. 2017. Available at: marketsandmarkets.com. Accessed on February 2017
[22] Lin LZ, Yao YG, Yoshioka M, Shiraishi N. Liquefaction mechanism of cellulose in the presence of phenol under acid catalysis. Carbohyd Polym. 2004;57:123–29.
[23] Lin L. Characterization of phenolated wood and study on the liquefaction mechanism of lignin. Kyoto: Kyoto University; 1996.
[24] Sturgeon MR, Kim S, Lawrence K, Paton RS, Chmely SC, Nimlos M et al. A mechanistic investigation of acid-catalyzed cleavage of aryl-ether linkages: implications for lignin depolymerization in acidic environments. ACS Sustainable Chem Eng. 2014;2:472–85.
[25] Zhao Y, Yan N, Feng MW. Effects of reaction conditions on phenol liquefaction of beetle-infested lodgepole pine barks. Curr Org Chem. 2013;17:1604–16.
[26] Zhao Y. Development of bio-based phenol formaldehyde resol resins using mountain pine beetle infested lodgepole pine barks [Thesis (Ph D)]: University of Toronto (Canada), 2013.
[27] Zhao Y, Yan N, Feng MW. Biobased phenol formaldehyde resins derived from beetle-infested pine barks structure and composition. ACS Sustainable Chem Eng. 2012;1:91–101.
[28] D'Souza J, Yan N. Producing bark-based polyols through liquefaction: effect of liquefaction temperature. ACS Sustainable Chem Eng. 2013;1:534–40.
[29] D'Souza J, Wong SZ, Camargo R, Yan N. Solvolytic liquefaction of bark: understanding the role of polyhydric alcohols and organic solvents on polyol characteristics. ACS Sustain Chem Eng. 2015;4:851–61.
[30] Ionescu M. Chemistry and Technology of Polyols for Polyurethanes, 2nd ed., vol. 2. Shrewsbury: iSmithers Rapra Publishing, 2016: 440.
[31] Nadji H, Bruzzese C, Belgacem MN, Benaboura A, Gandini A. Oxypropylation of lignins and preparation of rigid polyurethane foams from the ensuing polyols. Macromol Mater Eng. 2005;290:1009–16.
[32] Wu LCF, Glasser WG. Engineering plastics from Lignin 0.1. Synthesis of hydroxypropyl lignin. J Appl Polym Sci. 1984;29:1111–23.
[33] Cateto CA, Barreiro MF, Rodrigues AE, Belgacem MN. Optimization study of lignin oxypropylation in view of the preparation of polyurethane rigid foams. Ind Eng Chem Res. 2009;48:2583–89.
[34] D'Souza J, George B, Camargo R, Yan N. Synthesis and characterization of bio-polyols through the oxypropylation of bark and alkaline extracts of bark. Ind Crops Prod. 2015;76:1–11.
[35] Briones R, Serrano L, Ben Younes R, Mondragon I, Labidi J. Polyol production by chemical modification of date seeds. Ind Crops Prod. 2011;34:1035–40.
[36] Garcia DE, Glasser WG, Pizzi A, Osorio-Madrazo A, Laborie MP. Hydroxypropyl tannin derivatives from pinus pinaster (Ait.) bark. Ind Crops Prod. 2013;49:730–39.
[37] Pavier C, Gandini A. Oxypropylation of sugar beet pulp. 1. Optimisation of the reaction. Ind Crops Prod. 2000;12:1–8.
[38] Fernandes S, Freire CSR, Neto CP, Gandini A. The bulk oxypropylation of chitin and chitosan and the characterization of the ensuing polyols. Green Chem. 2008;10:93–97.

[39] Evtiouguina M, Gandini A, Neto CP, Belgacem NM. Urethanes and polyurethanes based on oxypropylated cork: 1. Appraisal and reactivity of products. Polym Int. 2001;50:1150–55.
[40] Ayers JW, Bradley JJJ. Phenolic resin glue compositions containing a vegetable shell flour derivative. In: Google Patents; 1957.
[41] Barrentine EM, Hodgins TS, Shelton FJ. Adhesive composition comprising phenol aldehyde resin and hydroxyalkylcellulose and method of preparing same. In: Google Patents; 1958.
[42] Duncan TF, Putnam RD. Adhesive composition comprising phenolic resin, a tannin and an alkali metal hydroxide. In: Google Patents; 1966.
[43] Hartman S. Alkali-treated bark extended tannin-aldehyde resinous adhesives. In: Google Patents; 1977.
[44] Kenneth RG, John CS, Hartzell LC. Bark treatment process and alkali bark derivative. In: Google Patents; 1959.
[45] Heritage CC, Dowd LE. Alkaline extraction of chemical products from bark. In: Google Patents; 1959.
[46] Baxter GF, Preusser HM, Schoening PK. Thixotropic phenol-aldehyde adhesive composition and method of making same. In: Google Patents; 1959.
[47] Heritage CC. Cork-bound hot-pressed boards. In: Google Patents; 1956.
[48] Feng SH, Cheng SN, Yuan ZS, Leitch M, Xu CB. Valorization of bark for chemicals and materials: A review. Renew Sust Energ Rev. 2013;26:560–78.
[49] Markets R. Global Phenolic Resin Market Analysis 2016 – Forecast to 2022. Dublin, In: Hoovers Research. 2017.
[50] Vazquez G, Antorrena G, Parajo JC, Francisco JL. Preparation of wood adhesives by polycondensation of phenolic-acids from pinus pinaster bark with resoles. Holz Roh Werkst. 1989;47:491–94.
[51] Fechtal M, Riedl B. Use of eucalyptus and acacia mollissima bark extract-formaldehyde adhesives in particleboard manufacture. Holzforschung Int J Biology, Chemistry, Phys Technol Wood. 1993;47(4):349.
[52] Zhao Y, Yan N, Feng MW. Bark extractives-based phenol–formaldehyde resins from beetle-infested lodgepole pine. J Adhes Sci Technol. 2013;27:2112–26.
[53] Zhao Y, Yan N, Feng MW. Thermal degradation characteristics of phenol–formaldehyde resins derived from beetle infested pine barks. Thermochimica Acta. 2013;555:46–52.
[54] Pizzi A. Advanced wood adhesives technology. New York: M. Dekker; 1994.
[55] Gao ZH, Yuan JL, Wang XM. Phenolated larch-bark formaldehyde adhesive with multiple additions of sodium hydroxide. Pigm Resin Technol. 2007;36:279–85.
[56] Alma MH, Acemioglu B. A kinetic study of sulfuric acid-catalyzed liquefaction of wood into phenol. Chem Eng Commun. 2004;191:968–80.
[57] Lee WJ, Liu CT. Preparation of liquefied bark-based resol resin and its application to particle board. J Appl Polym Sci. 2003;87:1837–41.
[58] Santana MD, Baumann MG, Conner AH. Utilization of Black Wattle Bark and Tannin Liquefied in Phenol in the Preparation of Resol- Type Adhesives In: Lignocellulosic-Plastics Composites. Leao A. Carvalho F. Frollini E. (ed.). 1997;325–341.
[59] Zhao Y, Zhang B, Yan N, Farnood R. Synthesis and characterization of phenol formaldehyde novolac resin derived from liquefied mountain pine beetle infested lodgepole pine barks. Macromol React Eng. 2013;7:646–60.
[60] Zhang B. Development of bio-based phenol formaldehyde novolac resins using mountain pine beetle infested lodgepole pine barks [Thesis (M A S)]: University of Toronto (Canada), 2014.
[61] Alma MH, Kelley SS, Bektas I. The tensile properties of molding products obtained by the condensation of various tree barks and phenol by using sulfuric acid as a catalyst. J Mater Sci Lett. 2000;19:1517–20.

[62] Chai Y, Zhao Y, Yan N. Synthesis and characterization of biobased melamine formaldehyde resins from bark extractives. Ind Eng Chem Res. 2014;53:11228–38.
[63] Kuo P-Y, Sain M, Yan N. Synthesis and characterization of an extractive-based bio-epoxy resin from beetle infested pinus contorta bark. Green Chem. 2014;16:3483–93.
[64] Kuo PY, Barros LD, Sheen YC, Sain M, Tjong JSY, Yan N. Thermal degradation of extractive-based bio-epoxy monomer and network: kinetics and mechanism. J Anal Appl Pyrol. 2016;117:199–213.
[65] Kuo P-Y, De Assis Barros L, Sain M, Tjong JS, Yan N. Effects of reaction parameters on the glycidyl etherification of bark extractives during bioepoxy resin synthesis. ACS Sustain Chem Eng. 2016;4:1016–24.
[66] D'Souza J, Camargo R, Yan N. Biomass liquefaction and alkoxylation: a review of structural characterization methods for bio-based polyols. Polymer Rev. 2017;57(4): 1–27.
[67] Zhao Y, Yan N, Feng M. Polyurethane foams derived from liquefied mountain pine beetle-infested barks. J Appl Polym Sci. 2012;123:2849–58.
[68] D'Souza J, Camargo R, Yan N. Polyurethane foams made from liquefied bark-based polyols. J Appl Polym Sci. 2014;131(16).
[69] Sankar G, Yan N. Bio-based two component (2 K) polyurethane adhesive derived from liquefied infested lodgepole pine barks. J Biobased Mater Bio. 2014;8:457–64.

Steven J. Bachofer and Mark D. Lingwood

8 A green determination of an equilibrium constant: teaching new skills

Abstract: The spectroscopic determination of an equilibrium constant is a classic experiment in the general chemistry laboratory curriculum. Here we describe a new version of this experiment, incorporating green chemistry practices into the general chemistry lab curriculum and directly teaching students about the principles of green chemistry. This lab studies the formation of the iron(III) salicylate complex ion, which is a more benign chemical system than the traditional iron(III) thiocyanate complex. A microtiter plate reader is used to reduce sample volumes, also giving students important experience using research-grade instrumentation. In addition, students gain more experience using adjustable manual micropipettors. Students responded positively to the lab, appreciating the green nature of the lab and the introduction of the plate reader, and student comments suggested that this lab successfully introduces green chemistry concepts to the general chemistry curriculum.

Keywords: Spectroscopic determination, waste reduction, iron(III) salicylate, plate reader, microtiter plate, manual pippettor

8.1 Introduction

Green chemistry principles are gaining more attention throughout the chemistry curriculum. Many professors are working to change their instructional labs to more effectively educate the next generation of chemists in green chemistry, as one aspect of teaching students of their ethical responsibilities as scientists [1]. For example, numerous organic chemistry labs have been developed through the GEMS project at University of Oregon [2], and a new lab manual was recently published which highlights the principles of green chemistry in adapting some rather traditional general chemistry labs [3].

In this work, we describe a new green chemistry lab for general chemistry, where the equilibrium constant of the iron(III) salicylate complex formation is determined spectroscopically. This lab fulfills the fundamental objectives of the General Chemistry

This article has previously been published in the journal Physical Sciences Reviews. Please cite as: Bachofer, S., Lingwood, M. A Green Determination of an Equilibrium Constant: Teaching new skills. *Physical Sciences Reviews* [Online] **2018**, *3*. DOI: 10.1515/psr-2017-0195

https://doi.org/10.1515/9783110562781-008

lab curriculum, and emphasizes green chemistry principles by clearly reducing the quantity of waste produced by the experiment and by using a more benign chemical system than the traditional experiment with the $[Fe(SCN)]^{2+}$ complex ion [4, 5]. This new experiment also introduces students to two standard biochemical tools that are typically incorporated into upper division chemistry and biochemistry labs: a microtiter plate reader is used to obtain the spectroscopic data, and manual micropipettes are used to create the solutions [6, 7].

The spectroscopic determination of an equilibrium constant has been a fundamental lab experiment for several decades. The classical example of the formation of the iron(III) thiocyanate complex has been used to instruct students on the method of continuous variation and applying Beer's law to compute an equilibrium constant [5]. It is a well-documented lab, yet the standard lab conditions involve dispensing reactant volumes up to 20.0 mL [5]. The $Fe(NO_3)_3$ and KSCN reagents are typically in the low millimolar concentrations; however, both are dissolved in 0.50 Molar (M) HNO_3 for controlling the pH, preventing precipitation of the iron, and maintaining a fixed ionic strength [8]. An institution can implement methods to reduce the sample volumes in numerous ways, yet the high acid concentration still makes this system less attractive in general. Other chemical equilibria are also available as documented, including equilibria that have stoichiometric relationships that are different than the simple 1:1 stoichiometry of the $Fe(SCN)^{+2}$ complex ion [9].

A greener complexation equilibrium is the formation of the iron(III) salicylate complex which matches the traditional $Fe(SCN)^{+2}$ equilibrium with a simple 1:1 stoichiometry. The iron(III) salicylate complex ion yields a purple colored solution and is easily quantified with absorbance measurements at 525 nm. The iron(III) salicylate complex under acidic conditions has been previously utilized to quantify salicylate in numerous products and reported in educational laboratory experiments [10]. This chemical system is greener in several ways. When compared to KSCN, the sodium salicylate appears to be less detrimental when released into the environment [11–13]. Also, the acid concentrations of the salicylate and iron solutions can be reduced, since the reagents are stable for at least a week at these low concentrations. The use of salicylate is further advantageous due to the connection with industrial chemistry and broader society: acetylsalicylic acid, better known as aspirin, is an enormously important chemical synthesized industrially on a large scale and consumed worldwide. Some sets of general chemistry laboratory experiments contain an experiment to synthesize aspirin from salicylic acid [14], which can be related to the experiment described here.

The iron(III) salicylate complex ion equilibrium is presented as simply the transition metal ion reacting with a ligand, salicylate to form the complex ion as shown in Figure 8.1.

However, with the added HCl concentration in the samples, two equilibria are acting simultaneously: the ionization of the salicylic acid and the complex ion formation as shown in Figure 8.2 and Figure 8.3. The equilibrium constant measured in this experiment encompasses both reaction steps. In our iteration of the experiment, the

Figure 8.1: The metal ligand complexation equilibrium presented to students in this lab.

multiple equilibria are not presented to students for simplicity, yet very competent students may recognize this aspect. In addition, changing the added HCl concentration will lead to changes in the magnitude of the experimentally determined complex ion formation constant.

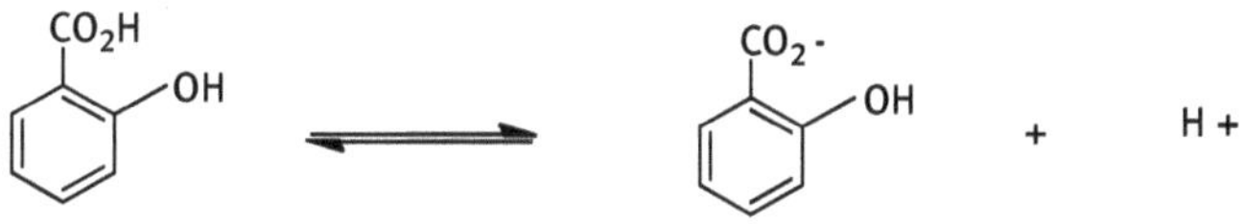

Figure 8.2: The acid ionization equilibrium of salicylic acid.

Figure 8.3: The resulting complexation equilibrium of iron (III) and salicylate anion.

The logic for having General Chemistry students utilize a plate reader to record absorbance measurements is to reduce laboratory waste, one of the goals of green chemistry. The smaller sample volumes required for the plate reader greatly contribute to the green nature of this experiment. The use of the plate reader in undergraduate labs is not necessarily novel, yet for the general chemistry lab curriculum it represents an advanced step. The educational values of incorporating plate readers into the curriculum are so well described by Santiago Botasini and coworkers that their article is a strongly suggested reading for anyone planning to incorporate this lab [15]. Very succinctly, Botasini and coworkers note that numerous biochemical experiments are typically recorded via plate reader formats, detecting absorption, fluorescence, or luminescence, so it is compelling to have students learn to use this type of instrument in a lower division course to prepare them for more advanced lab courses [15]. Furthermore, the authors highlight the utility of plate readers in clinical and industrial laboratories, therefore an early introduction to these instruments could aid students in gaining employment opportunities while they are pursuing their undergraduate degrees [15]. Finally, Botasini and coworkers performed a cost-benefit analysis and clearly demonstrate that one or even two lab sections can be supported by a single instrument due the rapid data collection speed of the instruments [15]. Our experience using plate readers is in agreement with this final point.

Another goal of this experiment is to give students more practice with adjustable manual micropipettors. This is not a novel element of this lab, as Nyasulu and Barlag recently presented the iron(III)thiocyanate equilibrium performed by small volume additions directly into the cuvette [16]. However for our college the task of micropipeting has become a standard practice for numerous upper division courses in chemistry and biology, so a few general chemistry labs introduce and reinforce proper pipetting technique. For our students, building proficiency in pipetting is a necessary skill. Also, in this lab the use of adjustable micropipettors facilitates the reduction in sample volumes, therefore reducing the quantity of waste generated. Furthermore, it would be difficult to precisely transfer the 0.2 to 1.5 mL volumes with other methods.

8.2 Learning goals

As noted above, the experiment fulfills both a traditional goal and introduces students to various aspects of green chemistry. The learning goals of this green determination of an equilibrium constant are as follows:

1. Students will compute an equilibrium constant value from experimental data.
2. Students will quantitatively use an adjustable micropipettor.
3. Students will record absorbance data using a microtiter plate and plate reader.
4. Students will note two or more principles of green chemistry.

While the primary learning goal is still for students to calculate an equilibrium constant from spectroscopic data, students utilize 96-well microtiter plates and plate reader to record the absorbance measurements and also use manual pipettors to make up quantitative samples. In this way, the students directly observe that the reagent volumes may be drastically reduced and will gain stronger pipetting skills to more adequately prepare them for other lab courses. The 96-well microtiter plates can be washed for reuse, so in addition to minimizing the volume of chemical used in the lab the plates can be reused without difficulties.

Furthermore, since the iron(III) salicylate does not completely eliminate the use of an acid to control the pH and electrolyte, the experiment allows students to consider methods to optimize the laboratory experiment further. In this way, the students can begin to internalize the principles of green chemistry with the intention that they may have additional novel ideas to improve the science that they are studying. Finally, students are directed to the Environmental Protection Agency's listing of the 12 principles of green chemistry [17, 18], and they are encouraged to consider what principles are highlighted in this experiment.

8.3 Details of experiment

This laboratory experiment is simple in design and it is frugal in materials used. The two reagents are 2.00×10^{-3} M ferric ammonium sulfate, $Fe(NH_4)(SO_4)_2$, and 2.00×10^{-3} M sodium salicylate, $NaC_6H_4(CO_2)(OH)$. Both reagents are prepared in 1.0×10^{-2} M hydrochloric acid, HCl, to yield a constant pH and nearly constant electrolyte concentration for the prepared samples. The maximum absorbance from the iron(III) salicylate complex ion demonstrates a 1:1 stoichiometry for the complex. With this knowledge, the experiment has students prepare samples very near the 1:1 stoichiometric ratio. Students prepare two sets of samples: ferric ammonium sulfate and sodium salicylate mixtures with some small variation in the stoichiometric ratio while holding the total volume constant; and mixtures with the two reagents set to 1:1 stoichiometry and adjusting the total volume using the 0.010 M HCl. In the first set of samples one trial is repeated to gauge the precision of students' sample preparation. For both sets of samples, the resulting equilibrium constant yields nearly the same numerical value. As noted in Table 8.1, the experimental volumes are sufficiently small that each pair of students needs no more than 10 mL of the two reagents and 10 mL of the 0.010 M HCl. The resulting iron(III) salicylate samples appear purple in color.

Table 8.1: Reagent sample volumes.

2.00×10^{-3} M ferric ammonium sulfate (mL)	2.00×10^{-3} M sodium salicylate (mL)	1.0×10^{-2} M hydrochloric acid (mL)	Total sample volume (mL)
0.45	0.55	1.00	2.00
0.50	0.50	1.00	2.00
0.50	0.50	1.00	2.00
0.55	0.45	1.00	2.00
0.50	0.50	1.50	2.50
0.50	0.50	0.80	1.80
0.50	0.50	0.50	1.50
0.50	0.50	0.30	1.30

The instructions direct students to mix the samples with a vortex mixer and then carefully transfer samples to the microtiter plate. For this experiment, students were instructed to use only flat bottom well plates made of polystyrene and directed to only transfer 0.20 mL using the appropriate pipettor. Students transferred each sample to three separate wells in the plate, acquiring absorbance data in triplicate. The experimental data were collected using a BioRad Benchmark plate reader. This instrument has a xenon flash lamp for the light source, and a monochromator with a range from 340 nm to 800 nm. The instrument can be configured to analyze various plates (including 6, 24, 96, 384 wells) and both endpoint and kinetic assays can be performed.

The instrument has a Peltier thermal control system (typically used for biochemical analysis) that was not needed for this experiment as samples were measured at room temperature (approximately 25 °C). The students are provided with the path length and an extinction coefficient for the iron(III) salicylate complex, which is the only species that has a measurable absorbance at 525 nm. The extinction coefficient was determined previously on the plate reader using standards, 2.00×10^{-3} M sodium salicylate and 1.00×10^{-1} M ferric ammonium sulfate, and delivering the same 0.20 mL volumes to microtiter plates with the same well shape. The extinction coefficient (2280 M^{-1} cm^{-1}) multiplied by the path length of 0.50 cm yielded a value of 1140 M^{-1}.

Typical absorbance values for the samples with a slight change in the mole fraction with a constant added HCl volume of 1.00 mL are 0.26. The 1:1 mole fraction samples at different total volumes had notably different absorbance values due to the different overall concentrations of reagents and products. Students used a spreadsheet program to calculate the K_{eq} value first by using Beer's Law to compute iron(salicylate) complex ion concentration, $[Fe(salicylate)]_{eq}$, then using the equilibrium concentration of the complex ion to calculate the equilibrium concentrations of the two reactants using the mass balance equation, and finally using the equation for the equilibrium constant to determine K_{eq}. The K_{eq} values under the conditions listed in Table 8.1 yielded a value of 3,300 M^{-1}. The values for the first four samples in the table (varying the stoichiometric ratio) were fairly constant between samples. The K_{eq} values for the last four samples in the table (varying overall volume and thus reactant concentration) decreased slightly as the overall reactant concentrations increased (i. e. volume of added HCl decreased). Experimental values were relatively similar between groups of students.

After the laboratory, the post-lab questions asked students to compare their results for the first four vs. the last four samples. Students were also asked to consider their relative levels of precision, in light of the replicate samples (second and third line in Table 8.1) and the absorbance measures in triplicate on each sample. Students were also directly asked how this lab demonstrates two or more principles of green chemistry [17] and were also asked to consider future ways this experiment could be made even greener. This last question was planned so that students will see themselves as individuals who can advocate for greening the lab curriculum in the future.

One obvious extension of this experiment would be to have students determine the extinction coefficient of the iron(III) salicylate complex. This could be accomplished on the same microtiter plate as their unknowns, by preparing a series of standards with iron concentrations 20 to 50 times higher than salicylate concentrations [10]. This lab could also be adapted to verify the 1:1 stoichiometry of the iron(III) salicylate complex, employing the method of continuous variation by measuring samples with a much wider range of reactant stoichiometric ratios and observing the stoichiometric ratio where the measured absorbance is at a maximum. In our implementation of the experiment, students were given lab time to process their data in a spreadsheet program, submitting the spreadsheet with their lab report.

8.4 Student response

To investigate the impact of this new green K_{eq} determination, an online survey was given to several lab sections of students after completing the lab and receiving grades on their laboratory notebooks and data. The Student Assessment of Learning Gains tool (SALG, www.salgsite.org) was utilized to survey the students. This assessment tool uses student self-reported data that is collected anonymously. This survey tool has a format to address what has helped the students to make gains. The student responses were on a traditional 1 to 5 scale (strongly disagree = 1 to strongly agree = 5). The SALG assessment tool also allows open-ended questions which provide additional insights. The students were asked numerous questions on what made the laboratory experiment instructional. In particular, questions focused on what using the plate reader meant in context to their collecting data and whether it was easy or difficult to collect the data using the 96-well microtiter plates. Plus, students were asked to respond to what green chemistry aspects were reinforced in this lab. The student responses in Table 8.2 demonstrate that many educational objectives are indeed reached.

Table 8.2: Aggregated student responses to SALG survey questions.

Questions *N* = 26 students Numerical responses	Rank 1 Strongly disagree	Rank 2 Disagree	Rank 3 Neutral	Rank 4 Agree	Rank 5 Strongly Agree	Mean
The 96-well plate was easy to use in preparing samples.	0	0	2	6	18	4.6
The absorbance measurements were easily obtained using the 96-well plate.	0	0	1	3	22	4.8
The volumes of reagents were reduced in this lab in comparison to a typical lab.	0	0	1	3	22	4.8
The rapid data collection allowed students to spend the appropriate time preparing the samples.	0	0	1	3	22	4.8
The pipettors were easy to use in preparing the samples.	0	0	2	5	19	4.7
The plate reader technology was straightforward to apply in this lab.	0	0	3	1	22	4.7

In open-ended questions, students were asked for their thoughts on what made the lab experiment green from their perspective. Two of the more illustrative responses which match numerous student comments were "we were more efficient with our time, use of reagents and energy usage" and "reducing hazardous substances made this lab a greener chemistry lab." In addition, when asked "the use of the plate readers was instructional because . . ." typical responses were "it allowed me to see how we can use less reagent and get even more accurate results due to the development of technology" and "it felt professional and we will most likely be using these in the future when conducting experiments."

8.5 Conclusion

In conclusion, this new general chemistry laboratory experiment accomplishes the desired educational objectives: students connect with the lecture course content by determining an equilibrium constant, they learn new skills while using the absorbance plate reader, and students bolster their skills using adjustable micropipettors. The concepts of green chemistry are successfully introduced to the students, who appear to appreciate the move toward smaller reaction volumes and less hazardous reagents. Students also appreciate the chance to work with research-grade instrumentation. This experiment empowers students to consider how the laboratory experiment could be improved with regard to the principles of green chemistry, and hopefully inspires students to advocate for green and sustainable practices in their future laboratory experiences.

Acknowledgements: The Chemistry Department began pursuing screening methods in the upper division curriculum with the generous support of the Thomas J. Long Foundation, who funded the purchase of the BioRad plate reader in 2006. This instrument is now utilized throughout the curriculum and in research. The additional efforts of Ian Ziraldo were also appreciated in the development of this laboratory experiment.

References

[1] Anastas PT, Kirchhoff MM. Origins, current status, and future challenges of green chemistry. Acc Chem Res. 2002 Sept;35(9):686–94. DOI: 10.1021/ar010065m.

[2] GEMS – Greener Education Materials for Chemists. [Internet]. Available at: http://greenchem.uoregon.edu/gems.html. Accessed: 30 May 2017.

[3] Henrie SA. Green chemistry laboratory manual for general chemistry. Boca Raton, FL: Taylor and Francis Group, CRC Press, 2015:213–28.

[4] Ramette RW. J Chem Educ. 1963 Feb;40(2):71–72. DOI: 10.1021/ed040p71.

[5] Postma JM, Roberts JL, Hollenberg JL. Chemistry in the laboratory, 5th ed. New York: Freeman, 2004:24–31. *This experiment remains in the current version of this text* (8th ed., *Macmillan Learning, 2017)*.

[6] Johnson RJ, Hoops GC, Savas CJ, Kartje Z, Lavis LD, Sensitive A. Robust enzyme kinetic experiment using microplates and fluorogenic ester substrates. J Chem Educ. 2015 Feb; 92(2):385–88. DOI: 10.1021/ed500452f.

[7] Marques JT, de Almeida RF. Application of ratiometric measurement and microplate fluorimetry to protein denaturation: an experiment for analytical and biochemistry students. J Chem Educ. 2013 Nov;90(11):1522–27. DOI: 10.1021/ed300599d.

[8] Cobb CL, Love GA. Iron(III)Thiocyanate revisited: a physical chemistry equilibrium lab incorporating ionic strength effects. J Chem Educ. 1998 Jan;75(1):90–92. DOI: 10.1021/ed075p90.

[9] Klotz E, Doyle R, Gross E, Mattson B. The equilibrium constant for bromothymol blue: a general chemistry laboratory experiment using spectroscopy. J Chem Educ. 2011 May;88(5):637–39. DOI: 10.1021/ed1007102.

[10] Mitchell-Koch JT, Reid KR, Meyerhoff ME. Salicylate detection by complexation with iron (III) and optical absorbance spectroscopy. J Chem Educ. 2008 Dec;85(12):1658–59. DOI: 10.1021/ed085p1658.

[11] Bhunia F, Saha N, Kaviraj A. Toxicity of thiocyanate to fish, plankton, worm, and aquatic ecosystem. Bull Environ Contam Toxicol. 2000 Nov;64:197–203. DOI: 10.1007/s001289910030.

[12] Gravel A, Vijayan MM. Non-steroidal anti-inflammatory drugs disrupt the heat shock response in rainbow trout. Aquat Toxicol. 2007;81(2):197–206.

[13] Zivna D, Sehonova P, Plhalova L, Marsalek P, Blahova J, Prokes M, et al. Effect of salicylic acid on early life stages of common carp (Cyprinus carpio). Environ Toxicol Pharmacol. 2015; 40(1):319–25.

[14] Brown TE, Nelson JH, Kemp KC. Laboratory experiments for chemistry: the central science, 13th ed. New York: Pearson; 2015.

[15] Botasini S, Lazuriaga L, Cerda MF, Mendez E. Multiple experiments and a single measurement: introducing microplate readers in the laboratory. J Chem Educ. 2010 Aug;87(8):1011–14. DOI: 10.1021/ed100789.

[16] Nyasulu F, Barlag R. Colorimetric determination of iron(III)-thiocyanate reaction equilibrium constant with calibration and equilibrium solutions prepared in a cuvette by sequential additions on one reagent to the other. J Chem Educ. 2011 Mar;88(3):313–14. DOI: 10.1021/ed1009927.

[17] U.S. EPA's Green Chemistry Principles. [Internet]. Available at: https://www.epa.gov/greenchemistry/basics-green-chemistry#twelve. Accessed: 1 June 2017.

[18] American Chemical Society summary of Green chemistry Principles. [Internet]. Available at: https://www.acs.org/content/acs/en/greenchemistry/what-is-green-chemistry/principles/12-principles-of-green-chemistry.html. Accessed: 1 June 2017.

Index

https://doi.org/10.1515/9783110562781-009

Also of interest

Volume 1
Sustainable Green Chemistry.
Benvenuto (Ed.), 2017
ISBN 978-3-11-044189-5, e-ISBN 978-3-11-043585-6
ISSN 2366-2115

Volume 2
Green Chemical Processes.
Developments in Research and Education
Benvenuto (Ed.), 2017
ISBN 978-3-11-044487-2, e-ISBN 978-3-11-044592-3
ISSN 2366-2115

GREEN - Alternative Energy Resources
Volume 1
Pyrolysis of Biomass.
Wang, Luo, 2016
ISBN 978-3-11-037457-5, e-ISBN 978-3-11-036966-3
ISSN 2509-7237

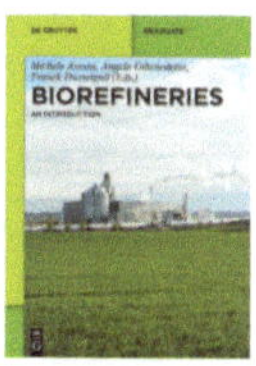

Biorefineries.
An Introduction
Aresta, Dibenedetto, Dumeignil (Eds.), 2015
ISBN 978-3-11-033153-0, e-ISBN 978-3-11-033158-5

Sustainable Process Engineering.
Prospects and Opportunities
Koltuniewicz, 2014
ISBN 978-3-11-030875-4, e-ISBN 978-3-11-030876-1

Bei Fragen zur Produktsicherheit wenden Sie sich bitte an:
If you have any questions regarding product safety,
please contact:

Walter de Gruyter GmbH
Genthiner Straße 13
10785 Berlin
productsafety@degruyterbrill.com